Toward a New Era in U.S. Manufacturing

The Need for a National Vision

Manufacturing Studies Board

Commission on Engineering and Technical Systems

National Research Council

NATIONAL ACADEMY PRESS

Washington, D.C. 1986

National Academy Press • 2101 Constitution Avenue, NW • Washington, DC 20418

NOTICE: The project that is the subject of this report was approved by the Governing Board of the National Research Council, whose members are drawn from the councils of the National Academy of Sciences, the National Academy of Engineering, and the Institute of Medicine. The members of the board responsible for the report were chosen for their special competences and with regard for appropriate balance.

This report has been reviewed by a group other than the authors according to procedures and approved by a Report Review Committee consisting of members of the National Academy of Sciences, the National Academy of Engineering, and the Institute of Medicine.

The National Research Council was established by the National Academy of Sciences in 1916 to associate the broad community of science and technology with the Academy's purposes of furthering knowledge and of advising the federal government. The Council operates in accordance with general policies determined by the Academy under the authority of its congressional charter of 1863, which establishes the Academy as a private, nonprofit, self-governing membership corporation. The Council has become the principal operating agency of both the National Academy of Sciences and the National Academy of Engineering in the conduct of their services to the government, the public, and the scientific and engineering communities. It is administered jointly by both Academies and the Institute of Medicine. The National Academy of Engineering and the Institute of Medicine were established in 1964 and 1970, respectively, under the charter of the National Academy of Sciences.

This work is related to Department of the Navy Grant N00014-85-G-0094 issued by the Office of Naval Research.

Library of Congress Catalog Card Number 86-50832

International Standard Book Number 0-309-03691-7

Printed in the United States of America

Manufacturing Studies Board

DAVID C. EVANS, President and Chairman of the Board, Evans and Sutherland Computer Corporation, Salt Lake City, Utah

W. PAUL FRECH,* President, Lockheed Georgia Company, Marietta, Georgia

BELA GOLD, Fletcher Jones Professor of Technology and Management, Claremont Graduate School of Business Administration, Claremont, California

DALE B. HARTMAN, Director of Manufacturing Technology, Hughes Aircraft Company, Los Angeles, California

MICHAEL HUMENIK, JR.,† Director, Manufacturing Process Laboratory, Ford Motor Company, Detroit, Michigan

ROBERT S. KAPLAN, Professor of Industrial Administration, Carnegie-Mellon University, Pittsburgh, Pennsyvlania, and Professor of Accounting, Graduate School of Business Administration, Harvard University, Boston, Massachusetts

JAMES F. LARDNER, Vice President, Component Group, Deere and Company, Moline, Illinois

M. EUGENE MERCHANT,* Director, Advanced Manufacturing Research, Metcut Research Associates, Inc., Cincinnati, Ohio

ROY MONTANA,‡ General Manager, Bethpage Operation Center, Grumman Aerospace Corporation, Bethpage, New York

THOMAS J. MURRIN, President, Energy and Advanced Technology Group, Westinghouse Electric Corporation, Pittsburgh, Pennsylvania

ROGER N. NAGEL, Director, Manufacturing Systems Engineering, Lehigh University, Bethlehem, Pennsylvania

PETER G. PETERSON, Peterson, Jacobs & Co., New York, New York

RAJ REDDY, Director, Robotics Institute, and Professor of Computer Science, Carnegie-Mellon University, Pittsburgh, Pennsylvania

DAN L. SHUNK, Director, Center for Automated Engineering and Robotics, Arizona State University, Tempe, Arizona

WICKHAM SKINNER,* Graduate School of Business Administration, Harvard University, Boston, Massachusetts

*Term expired before project completion
† Deceased
‡ Resigned

Staff of the Manufacturing Studies Board

GEORGE H. KUPER, *Executive Director*
JANICE E. GREENE, *Staff Officer*
GEORGE D. KRUMBHAAR, JR.,* *Staff Officer*
THOMAS C. MAHONEY, *Staff Officer*
MICHAEL A. MCDERMOTT, *Staff Officer*
DENNIS A. DRISCOLL, *Staff Associate*
LUCY V. FUSCO, *Staff Assistant*
DONNA REIFSNIDER,† *Staff Assistant*
MICHAEL S. RESNICK, *Staff Assistant*
CAROLYN CASTORE, *Staff Officer*: Committee on the Role of the Manufacturing Technology Program in the Defense Industrial Base (on Intergovernmental Personnel Administrative Exchange from the General Accounting Office)
GERALD I. SUSMAN, *Project Director*: Committee on the Effective Implementation of Advanced Manufacturing Technologies (on leave from The Pennsylvania State University through December 1985)
MARGARET DEWAR, National Research Council Fellow on leave from the Humphrey Institute, University of Minnesota through August 1985

*Employed through August 31, 1985
†Employed through May 17, 1985

Preface

U.S. manufacturing has entered a new era, created by the convergence of three important trends:

- the rapid spread of manufacturing capabilities worldwide;
- the emergence of advanced manufacturing technologies; and
- growing evidence that appropriate changes in traditional management and labor practices and organizational structures are needed to improve the competitiveness of U.S. manufacturing operations.

The responses of U.S. manufacturers to these trends will determine their long-term competitiveness and the future prosperity of the U.S. economy.

Relatively few domestic manufacturers have devised effective responses to ensure success in the new manufacturing environment. Despite encouraging signs in certain industries during the recent economic recovery, the challenge from foreign manufacturers has continued to grow stronger in both the United States and foreign markets, the rate of investment in U.S. manufacturing remains disappointing, and the use of advanced technologies—which U.S. manufacturers claim as a competitive advantage—and new managerial practices is at best only comparable to their use by our competitors abroad. Unless U.S. manufacturers make progress soon, implementation of new practices, technologies, and strategies may be too slow, risking continued declines in competitiveness.

One aspect of the problem is that information on new developments in manufacturing is not sufficiently available to the manufacturing and policymaking communities. Recognizing this need, the Manufacturing Studies Board (MSB) of the National Research Council has produced this vision of the developments that are shaping the new manufacturing environment. The Board comprises experts from industry, labor, and academe who have had direct experience in the development and implementation of advanced manufacturing technologies. This report draws on current research, personal experience, and knowledge gained from several past MSB studies on specific developments in manufacturing.

The report presents a broad view of what the MSB believes will be necessary to maintain a competitive U.S. manufacturing sector and some of the major considerations manufacturers and government policymakers will face in implementing these approaches. A general report describing the future manufacturing environment and emphasizing the manufacture of discrete parts was determined to be the best way to represent realistically the opportunities and issues confronting most manufacturers and government policymakers. Although the vision presented here is just one of many possibilities and is conditional on many factors in many fields, it is an optimistic scenario that effectively illustrates many of the relevant issues.

A major difficulty in presenting a report of this nature is the diversity of manufacturing. The technological needs, challenges to management and labor, and competitive situations of the many industries that constitute U.S. manufacturing are quite different. This diversity makes it virtually impossible to provide specific "how to" recommendations for either the private or public sector. Consequently, the report does not present an implementation manual or specific policy recommendations, but it does indicate a need for change and areas that should be addressed.

The report originated as a technology forecast, but through continued discussion the Board realized the inadequacy of such an approach. Recognizing that human resource and management issues will be far more important to future manufacturing competitiveness, the MSB has tried to emphasize that, despite the enthusiastic claims of technology developers and vendors, technology alone will not improve competitiveness. Without changes in corporate culture, organizational structures, and human resource management, new technologies will not produce the results needed for competitive manufacturing. These changes are far more difficult than plugging in a new machine—they require creative thinking, new attitudes, and a willingness to embrace change. These factors represent the critical barriers to improved competitiveness.

These adjustments will not be made simply or quickly, but their benefits can be profound. This report presents a vision of what those benefits can be and the imperative need to realize them. In writing it, the MSB hopes to make knowledge of current and future developments more available, to stimulate national discussion of the related issues, and to help ensure that U.S. industry is a world leader in the new manufacturing era.

Robert B. Kurtz, *Chairman*
Manufacturing Studies Board

Acknowledgments

This report was conceived and produced by the Manufacturing Studies Board over the past 2 years. It was funded by a grant from the U.S. Departments of the Army, Navy, and Air Force, administered by the Office of Naval Research under the direction of the Under Secretary of Defense for Research and Engineering.

As with any work of this kind, the report represents the contribution of many more individuals and organizations than can be acknowledged in the available space. The Board would like to express its appreciation to all those who have been involved directly and indirectly in what it believes is a consensual process of direction-setting for U.S. manufacturing.

A number of individuals deserve special recognition. Erich Bloch, former chairman of the Manufacturing Studies Board and current director of the National Science Foundation, was instrumental in the genesis of the project. John Lyons, John McTague, D. Bruce Merrifield, John Mittino, Everette Pyatt, Donald Rheem, William Schmidt, L. William Seidman, and James Spates offered their help and insight during the early stages of the project. Others contributed their time and energy to the review and refinement of the report in its various stages of evolution. Gerald Susman and Margaret Dewar reviewed early drafts, providing invaluable critique and insight. Keith McKee, Wickham Skinner, John Dunlop, John White, Ralph Gomory, George Heilmeier, and Solomon Buchsbaum greatly strengthened the report through their written reviews and suggestions. Many others, including Reginald Jones, James Baughman, George Carter, and Louis Cabot, provided the benefit of their experience and insight. The Board extends its thanks to these individuals and the many others who contributed so greatly to the substantive understanding that is reflected in the report and the bibliography.

Finally, the Board is particularly grateful to the National Research Council staff. Members of the Manufacturing Studies Board wrote and rewrote innumerable drafts, but the final text was crafted by Staff Officer Thomas Mahoney, whose patience, perseverance, dedication, and insight in this task deserve much of the credit for the clear presentation of a number of complicated ideas. Executive Director George Kuper guided and pushed the progress of the report and contributed many of the ideas at its heart. Staff Officers Janice Greene and George Krumbhaar helped draft the many early versions of the report, and Staff Associate Dennis Driscoll conducted much of the research on economic data. Consultants Roger Wright, Scott Garrigan, Catherine Rusinko, and Louis Blair provided valuable support in drafting and reviewing critical elements of the report. Edgar Weinberg conducted the research and wrote Appendix C, which summarizes past reports on U.S. manufacturing, shortly before his untimely death. Ronald Cowen and Kenneth Reese edited the final draft.

All this help and support notwithstanding, this report is the product of a hardworking and dedicated group of individuals. It has been my privilege to be their chairman at a time of such personal sacrifice and contribution.

Robert B. Kurtz

Contents

APPENDIXES

Toward a New Era in U.S. Manufacturing

Executive Summary

Manufacturing has entered the early stages of a revolutionary period caused by the convergence of three powerful trends:

- The rapid advancement and spread of manufacturing capabilities worldwide has created intense competition on a global scale.
- The emergence of advanced manufacturing technologies is dramatically changing both the products and processes of modern manufacturing.
- Changes in traditional management and labor practices, organizational structures, and decision-making criteria represent new sources of competitiveness and introduce new strategic opportunities.

These trends are interrelated and their effects are already being felt by the U.S. manufacturing community. Future competitiveness for manufacturers worldwide will depend on their response to these trends.

Based on the recent performance of U.S. manufacturers, efforts to respond to the challenges posed by new competition, technology, and managerial opportunities have been slow and inadequate. Domestic markets that were once secure have been assailed by a growing number of foreign competitors producing high-quality goods at low prices. In a number of areas, such as employment, capacity utilization, research and development expenditures, and capital investment, trends in U.S. manufacturing over the last decade have been unfavorable or have not kept pace with major foreign competitors. There is substantial evidence that many U.S. manufacturers have neglected the manufacturing function, have overemphasized product development at the expense of process improvements, and have not begun to make the adjustments that will be necessary to be competitive.

These adjustments represent fundamental changes in the way U.S. manufacturers perceive their competitive advantages, devise competitive strategies, and manage and organize their operations. One response that is beginning to gather momentum among U.S. manufacturers is the implementation of advanced manufacturing technology. Indeed, technology, wisely applied, can improve costs, quality, flexibility, and responsiveness, but the effects of technology on these areas can be complex. Trade-offs between improving flexibility and responsiveness on the one hand and reducing costs on the other will continue; technologies that are poorly applied may not have the effects intended; and many barriers to their smooth operation remain. Effective implementation of new technology demands a clear definition of the business's strategy and a clear understanding of the role of advanced technologies in supporting that strategy. Managers must recognize that many of the perceived advantages of new technology can be achieved with new management techniques, more effective planning, better coordination across corporate functions, efforts to reduce set-up times and speed changeovers, and simplified part designs to enhance producibility. Having made effective operational and organizational changes, the company can eliminate many of the problems that are often associated with the introduction of new technologies. Effective efforts in these areas also should help managers focus new investments on appropriate technology that can produce dramatic benefits.

These required organizational changes, however, will be difficult for many manufacturers to implement. They require creative initiatives from managers, the cooperation and involvement of employees, and major changes in the relationships at every level of the manufacturing corporation. A fundamental cultural and attitudinal shift will be required on the part of both workers and managers. Manufacturing will need to be thought of as a system, with extensive integration, cooperation, and coordination between functions, to achieve competitive goals. Flatter organizational structures are likely to become the norm and traditional hierarchical relationships are likely to fade as the distinctions between managers and workers blur. Workers will have more responsibility and greater job security and be more active participants in the manufacturing system.

Because the successful implementation of this cultural revolution in the factory depends on thousands of individual initiatives, change is likely to be gradual. In many cases, there will be strong resistance from both managers and workers who have a stake in traditional practices and structures. However, as competition in the new environment intensifies and the requirements to maintain competitive advantage with quality manufactured goods become clear, the benefits and the necessity of implementing these changes will be increasingly apparent.

These changes imply that the factory will provide a much different working environment and play a different role in the macroeconomy. For example, manufacturing will provide fewer job opportunities for unskilled and semiskilled workers, but the jobs that will be created are expected to require greater amounts of skill and training and thus to be more challenging and rewarding. Many manufacturers will have sufficient flexibility built into their production processes to be less affected by shifts in demand, which could moderate business cycles substantially.

These and other effects will require that government officials and the general public adjust their image and expectations of manufacturing. Although the technological and managerial changes necessary for future competitiveness will be the responsibility of the private sector, the government can play an important role in encouraging and supporting these private initiatives. Policymakers must recognize the continuing importance of manufacturing, the need for changes to ensure future competitiveness, and the many repercussions government policies have on the ability of U.S. manufacturers to meet competitive challenges. In addition, some specific government activities, in trade, education, research, and defense, will be affected by developments in manufacturing and will need to adapt accordingly.

For both government and industry, circumstances will vary tremendously. It is not possible to predict the specific strategies, technologies, management practices, and policies that will be effective in every situation. Instead, the Manufacturing Studies Board has described likely developments in the technology and human resource practices of future competitive manufacturers, as well as likely repercussions for government policies. By defining the direction in which U.S. manufacturers will need to move, and by raising some of the issues that they are likely to confront, the Board hopes to stimulate debate and involvement of a broad talent base on an opportunity of major national significance: accelerating the changes necessary to maintain the competitiveness of future U.S. manufacturing.

1
Statement of the Problem

For U.S. manufacturing, an extended period of world dominance in manufacturing innovation, process engineering, productivity, and market share has ended. Other countries have become leaders in certain industries, the U.S. market is being flooded by manufactured imports, and U.S. manufacturers are faced with relatively low levels of capacity utilization and declining employment. The reasons for this fundamental change are complex. Improved capabilities and competence of foreign manufacturers are partly responsible. Either government interference or the lack of government support has been blamed. Cultural disadvantages are often cited. Many economists explain the relative decline of U.S. manufacturing simply as economic evolution, with the United States moving toward a service economy. These and other factors have been held responsible for the relative decline of U.S. manufacturing, and all are legitimate partial explanations. The truth remains, however, that U.S. manufacturing is not performing as well as that of many foreign competitors and has lost competitiveness in many industries. Regardless of why the environment has changed, the managerial practices, strategies, and organizational designs applied by U.S. manufacturers have not adapted sufficiently to the changed competitive environment, and, consequently, U.S. manufacturing has not been as successful as that of other countries.*

These changes in relative manufacturing strength are occurring at the same time that many technological innovations promise to revolutionize products and processes in manufacturing. Just as major technological breakthroughs spurred industrial development in the mid-eighteenth century (steam power, new engine-driven machinery) and the development of the modern factory system in the late nineteenth century (electricity, the telephone, and mass production techniques), current breakthroughs in electronics, materials, and communications are creating another revolution in manufacturing. Just as earlier changes forced new directions in manufacturing management, production strategies, and national policies for maximizing competitiveness, the competitive and technological changes affecting manufacturing today should create new goals, new priorities, and new expectations in U.S. industry. Many manufacturing managers and national policymakers, however, have been slow to recognize the implications of these developments. U.S. manufacturing is in danger of being unpre-

---

* The term *competitiveness* is subject to a variety of definitions. In simplest form, an industry is competitive if the price, quality, and performance of its products equal or exceed that of competitors and provide the combination demanded by customers. International competitiveness is somewhat more complicated because price is heavily influenced by exchange rates, which cannot be controlled by an individual producer. Many economists would claim that the recent high rate of the dollar has been responsible for any lost competitiveness of U.S. manufacturing, and recent adjustments to the dollar will restore competitiveness. This may or may not be true, however, because exchange rates are only one determinant of product price, and price is only one determinant of competitiveness. Price is also determined by production costs, and quality and performance, including innovation, unique or superior design, and reliability, are in many cases more important determinants of competitiveness than price. If U.S. manufacturers can produce high-quality goods with less labor, materials, overhead, and inventory than foreign producers, then competitive production can be ensured. These are the areas in which U.S. manufacturers have lagged—improvements in the use of these resources, as well as product quality and performance, are the requirements for improved competitiveness.

pared to compete in the coming age, a failure that would cause rapid erosion of the nation's manufacturing base.

Effective response to the changes in manufacturing depends on a clear understanding of the new environment. Although specific developments are difficult to predict with certainty and the types of changes will vary tremendously among industries, likely trends can be identified. Competition will continue to increase both at home and abroad. New products will proliferate; many products will have shorter life cycles and development cycles. Some industries will have smaller production volumes, with more product customization and variety. New technologies, especially those based on microprocessors, will optimize control of the production process and offer entirely new capabilities. Fewer production workers and middle managers will be needed, but the remaining jobs will require higher skill, more technical knowledge, and greater responsibility. Managers will need to manage manufacturing as a system, basing decisions on new, nontraditional indicators. Direct labor costs will decrease significantly, and the costs of equipment, materials, distribution, energy, and other overhead will grow in importance.[1] Quality, service, and reliability will receive much more emphasis as determinants of competitive production.

These trends indicate that competition, both international and domestic, will be more intense and that the factors determining competitiveness will differ substantially from past experience. Strategies and priorities designed to enhance competitiveness in the mid-twentieth century will be far less effective in the future. The new manufacturing environment will be sufficiently familiar to permit many firms to continue to use traditional approaches, but these firms will lose market share, profits, and the ability to compete. In the new environment, it will not be sufficient to do the same old things better. Companies will need to adopt new management techniques, organizational structures, and operational procedures to strengthen their international competitiveness. Government policies must also ensure that U.S. manufacturers receive the infrastructural support they will need to compete effectively.

TABLE 1 Changes in U.S. Manufacturing Output[a]

Period	Average Annual Percentage Change: Total	Durable Goods	Nondurable Goods	As a Percentage of Total Output[b] (average)
1950-1983	3.1	3.0	3.1	24.4
1950-1973	4.0	4.0	4.0	24.6
1973-1983	0.9	0.7	1.1	24.1
Slowdown	3.1	3.3	2.9	0.5

[a]Gross product originating in manufacturing in constant dollars.
[b]Gross national product in constant dollars.

SOURCE: U.S. Bureau of Labor Statistics, 1985.

A HISTORICAL PERSPECTIVE ON U.S. MANUFACTURING

For much of the twentieth century, U.S. manufacturers were unchallenged in an environment in which conservative approaches to both process technology and managerial techniques produced successful results. Foreign competition was minimal, the vast domestic market encouraged product standardization and economies of scale, and the preeminence of Yankee ingenuity was unchallenged. Companies modified strategies and processes in minor ways in response to shifting economic circumstances, but mostly the system worked and they had little incentive to change. The relative stability of the manufacturing environment was unsustainable, however; a series of changes has gradually converted the traditional strategies to handicaps.[2]

One change has been in the way companies justify new investment in manufacturing. During the 1950s and 1960s, the emphasis in manufacturing was on providing substantial additional plant capacity that was needed just to keep up with market growth. The addition of capacity provided the opportunity to incorporate process improvements that otherwise were rarely implemented. Beginning in the early 1970s, the rate of growth slowed (Table 1), in many cases eliminating the need for additional capacity. Companies needed to develop new justifications for reinvestment in manufacturing, which many have been slow to do.[3]

Another major change in the manufacturing environment was in the process of developing and implementing new innovations. The first Industrial Revolution in the 1800s produced a series of significant innovations in process and product technologies that represented an integration of several types of technologies. In contrast, during the early to mid-1900s, manufacturers, except perhaps for electronics and chemicals manufacturers, increasingly refined proven technologies rather than developing and integrating new and diverse technologies to accomplish, or even eliminate, traditional tasks. This apparent trend toward a more stable, conservative approach to process technology in a broad range of U.S. industries combined with a variety of other factors—such as changing labor demographics, higher energy prices, and lower expenditures on research and development—to cause a shift toward more modest improvements in productivity.

U.S. industries in which new technology did seem to offer great potential focused predominantly on product engineering at the expense of process engineering. (The semiconductor, chemical, and biotechnology industries are exceptions—most of the breakthroughs in their products depend on breakthroughs in process capabilities.) Since manufacturers had their hands full simply adding capacity of a known type, they saw no pressing need to add new process technologies at the same time. Consequently, many U.S. firms spent incremental dollars on product technology and very little on new process technology. Generally speaking, U.S. manufacturers left process development to equipment suppliers and allowed their own skills at such development—and its link with product technologies and product quality—to decline.

THE CURRENT ROLE OF THE MANUFACTURING FUNCTION

These historic trends illustrate aspects of the manufacturing environment that have shaped the strategies of U.S. managers. For these and a variety of other reasons rooted in the history of industrial development,[4] many managers have focused on increasing the productivity of the manufacturing function, emphasizing production volume instead of product quality and reliability and process development. They believe that manufacturing, at best, can simply provide adequate support for competitive advantages in marketing or design engineering. It is true that many firms, particularly those in the *Fortune* 500, do enjoy substantial advantages in manufacturing owing to economies of scale and degrees of specialization that they have been able to achieve as large organizations. Generally, however, the charge to manufacturing, even in these companies, has been "Make the product—without any surprises."[5]

The traditional view in many U.S. firms is that manufacturing is a problem that can be solved with a given process at a given time. That process is then operated efficiently, with little incremental upgrading, until a significant improvement or new technology is implemented by competitors. This command-and-control view of manufacturing is based on the premise that smart people should be able to determine the optimal solution (process) for handling the tasks of the manufacturing function and then control the process and organization for maximum stability and efficiency until some external event forces change. Since the time between changes varies, the repercussions of this view may not be readily apparent. The key point is that it is a reactive view that overlooks the potential contributions of the manufacturing function to overall competitiveness.

Such an approach can erode the strength and competitive advantage provided by manufacturing. Quality, reliability, and delivery problems get blamed on manufacturing—the plan is assumed to be good, so the people in production must have failed to deliver. The organization increasingly refines the detailed measures of manufacturing in the process of removing degrees of freedom. Scientific management techniques were developed to measure, predict, and control all the aspects of production in an effort to limit change, or at least eliminate surprises, and achieve maximum productivity. Advances in production planning, project evaluation, and operations research offered new tools for maintaining stability and increasing productivity. The introduction of computers and manufacturing information systems in the late 1960s and early

1970s was hailed as finally giving manufacturing a tool that could be used to pursue the command-and-control approach to operations. Although designed to ensure stability in daily operations, these detailed measurements and sophisticated control tools too often can become ends in themselves and impediments to process changes.

The consequence of this approach to manufacturing has been increased tuning and refining of a set of resources that were outdated and increasingly inappropriate. The individual firm often slipped into a debilitating spiral: additional investment was withheld because the current investment was not performing as expected; those operating the current investment simply tried to minimize the problem in the near term rather than looking for long-term solutions they knew would not be approved and supported.

RECENT PERFORMANCE OF U.S. MANUFACTURING

The repercussions of this command-and-control approach, with its reactive nature and short-term focus, are not difficult to find. The United States has sustained a steady erosion of competitiveness and overall manufacturing strength over the past two decades that must be attributed at least partly to deficiencies in standard management practices in manufacturing. Individual companies have adapted to the new environment and fared well, but overall the picture has been bleak. Declining growth trends in manufacturing output have already been cited. Other indicators include

- Growth in manufacturing productivity (output per manhour) in the United States during the past 25 years has been among the lowest in the industrial world (Table 2). Although manufacturing productivity in this country remains the world's highest, it has been virtually equaled in recent years by Japan, France, and West Germany. Based on average hourly compensation and output per hour, unit labor costs in U.S. manufacturing have been higher than those of our major competitors (Table 3).
- In contrast to the growth in manufacturing trade surpluses enjoyed by Japan and West Germany, U.S. performance over the

TABLE 2 Output per Hour in Manufacturing, Average Annual Percent Change

Country	1960-1973	1973-1983
Canada	4.7	1.8
France	6.5	4.6
West Germany	5.7	3.7
Japan	10.5	6.8
United Kingdom	4.3	2.4
United States	3.4	1.8

SOURCE: U.S. Bureau of Labor Statistics, 1985.

past 15 years has been highly erratic, with deficits of $31 billion in 1983, $87 billion in 1984, and more than $100 billion in 1985 (Table 4).

- By 1984, manufacturing output was 8 percent above the previous peak in 1979. Defense production, however, accounted for more than 40 percent of that increase; nondefense output has risen less than 1 percent annually since 1979, compared with 3.5 percent annually from 1973 to 1979.[6]
- Recent employment trends have been unfavorable in most durable goods manufacturing industries, particularly import-competing industries (Figure 1).
- Capital investment as a percentage of output in U.S. manufacturing has increased slightly over the past 10 years (Table 5), but the composition of investment has tended to neglect traditional industries and new factory construction.[7] Although U.S. manufacturing investment has shown some improvement, it has continued to be below that in other countries.
- A major reason for the level and types of investment in manufacturing, cited by other reports on U.S. manufacturing,[8] is the high cost of investment capital. The cost of capital in this country is far higher than in other nations, and the return on manufacturing assets has not kept pace with the return on financial instruments (Table 6, Figure 2). In addition to the obvious impact this differential has on investment costs, lower capital costs and different sources of capital allow some foreign competitors to succeed with much lower rates of after-tax profit on sales than

TABLE 3 Foreign Labor Cost Components in Relation to U.S. Producers[a]

Country	1970	1975	1980	1981	1982	1983	1984
			Average Hourly Compensation				
France	41	72	92	75	68	63	56
West Germany	56	97	125	97	90	85	75
Italy	42	73	81	68	63	62	58
Japan	24	48	57	57	49	50	50
Korea	N.A.	6	11	10	10	10	10
United Kingdom	36	51	75	65	58	51	46
United States	100	100	100	100	100	100	100
			Output per Hour				
France	65	70	82	81	85	86	87
West Germany	66	71	79	78	79	79	78
Italy	56	60	70	70	71	68	69
Japan	44	52	72	74	79	79	84
Korea	N.A.	15	17	18	18	17	18
United Kingdom	41	43	42	43	44	45	44
United States	100	100	100	100	100	100	100
			Unit Labor Costs				
France	63	103	112	92	80	74	65
West Germany	85	136	157	123	114	107	95
Italy	75	123	115	96	89	91	84
Japan	53	92	79	77	62	63	60
Korea	N.A.	39	63	57	59	60	56
United Kingdom	86	120	177	151	132	115	105
United States	100	100	100	100	100	100	100

NOTE: N.A. = not available.

[a] Based on U.S. dollar values derived from average annual exchange rates.

SOURCE: Data Resources, Inc., 1985.

U.S. firms (1-2 percent versus 5-6 percent for U.S. firms). This difference effectively provides extra funds for capital investment or research and development expenditures.

There are a myriad of explanations for these troubling trends in U.S. manufacturing. Management, labor, and government all

TABLE 4 Trade Balance in Manufacturing, Billions of U.S. Dollars

Year	United States	Japan	West Germany
1970	3.4	12.5	13.3
1971	0.0	17.1	15.0
1972	−4.0	20.3	17.7
1973	−0.3	23.3	28.7
1974	8.3	38.0	42.4
1975	19.9	41.7	38.7
1976	12.5	51.2	42.1
1977	3.6	63.0	46.9
1978	−5.8	74.2	53.5
1979	4.5	72.0	59.2
1980	18.8	93.7	63.1
1981	11.8	115.6	61.7
1982	−4.3	104.0	67.5
1983	−31.0	110.3	58.7
1984	−87.4	127.9	60.5
1985	−107.5	107.7	59.5

SOURCE: U.S. Department of Commerce, Bureau of Economic Analysis, 1985, and International Trade Administration, 1986.

share responsibility. Macroeconomic factors such as domestic interest rates, exchange rates, the availability and cost of labor, foreign and domestic trade policies, and the constant seesaw of business cycles all have had an impact. Uncertainty about government spending, tax, and regulatory policies, and changes in the relative attractiveness of nontechnological (even nonmanufacturing) investments have deterred risky investments in new process technologies and bred caution in managers. Pressure from stockholders, standard financial evaluation procedures, and the disruptive effect that new technology can have on short-term operational efficiency also have caused managers to give priority to maximizing returns on existing assets.[9]

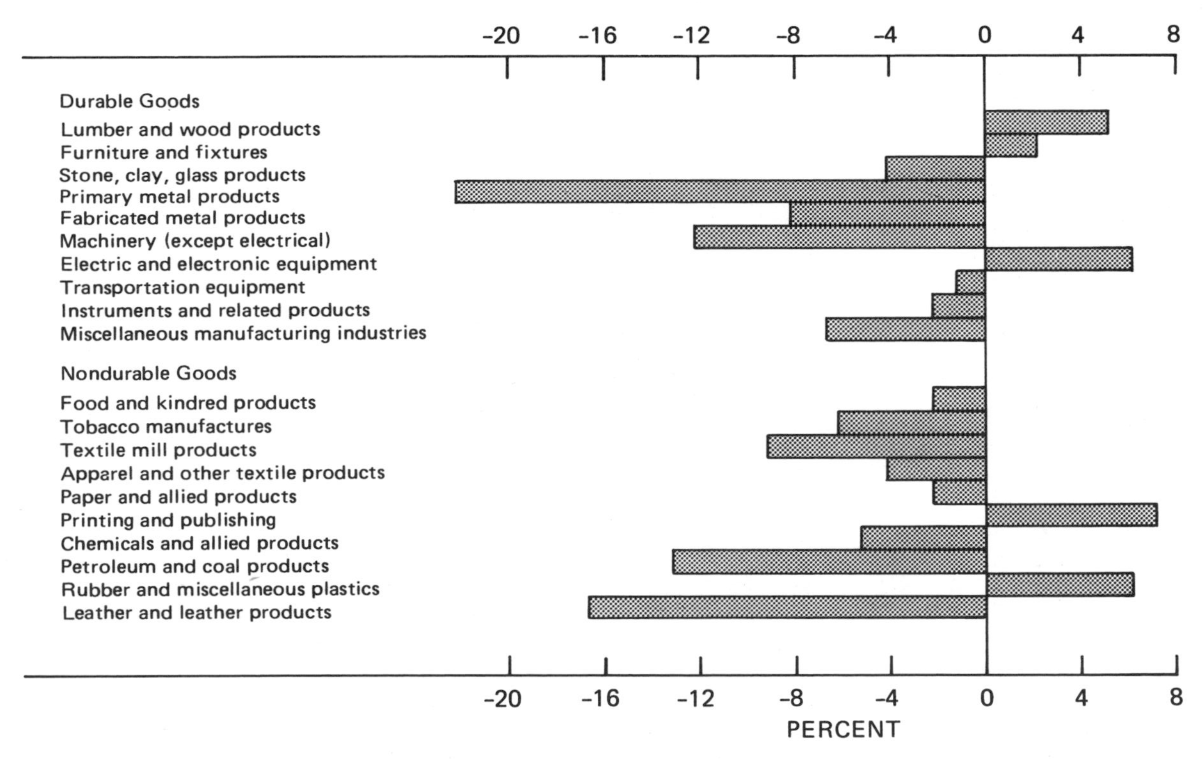

FIGURE 1 Change in manufacturing employment, selected industries (July 1981 to May 1984). Based on seasonally adjusted data for private manufacturing; includes only payroll employees. Source: U.S. Department of Labor, Bureau of Labor Statistics, establishment survey data.

TABLE 5 Capital Investment as Percentage of Output[a] in Manufacturing, Selected Countries, 1965-1982

Period	France	West Germany	Japan	United Kingdom	United States
1965-1982	15.1	12.8[b]	21.2	13.6	10.5
1965-1973	16.5	14.3	25.3	14.3	10.0
1974-1982	13.6	11.2[c]	17.1	13.0	11.1

[a]Fixed capital and output measured in constant dollars.
[b]1965-1981.
[c]1974-1981.

SOURCE: U.S. Bureau of Labor Statistics, 1985.

OTHER EVIDENCE OF A CHANGED MANUFACTURING ENVIRONMENT

Because of the diversity of the manufacturing sector and the factors affecting manufacturing output and trade, there is little agreement among economists and policymakers that U.S. manufacturing is losing competitiveness. Some authors have used economic data to demonstrate that U.S. manufacturing remains generally strong despite the problems of a few industries.[10] Many reports have addressed the issue by using macroeconomic data, but they have had little impact on either policymakers or the general public.

Statistics on the manufacturing sector tend to be inconclusive because of the complex, transient economic factors that affect

TABLE 6 Average Weighted Cost of Capital to Industry, 1971-1981 (in percent)

Country	1971	1976	1981
France	8.5	9.4	14.3
West Germany	6.9	6.6	9.5
Japan	7.3	8.9	9.2
United States	10.0	11.3	16.6

SOURCE: U.S. Department of Commerce, "Historical Comparison of Cost of Capital," April 1983.

the data. Other indicators, however, show that at least some U.S. companies have perceived both eroding competitiveness and a basic change in the nature of the manufacturing environment. These data tend to be anecdotal and industry specific and can be illustrated by a few examples.

- Through improved management, changed work rules, large investments in automation, and a variety of other measures, the three major U.S. automobile manufacturers have reduced their break-even volume for domestic production by more than 30 per-

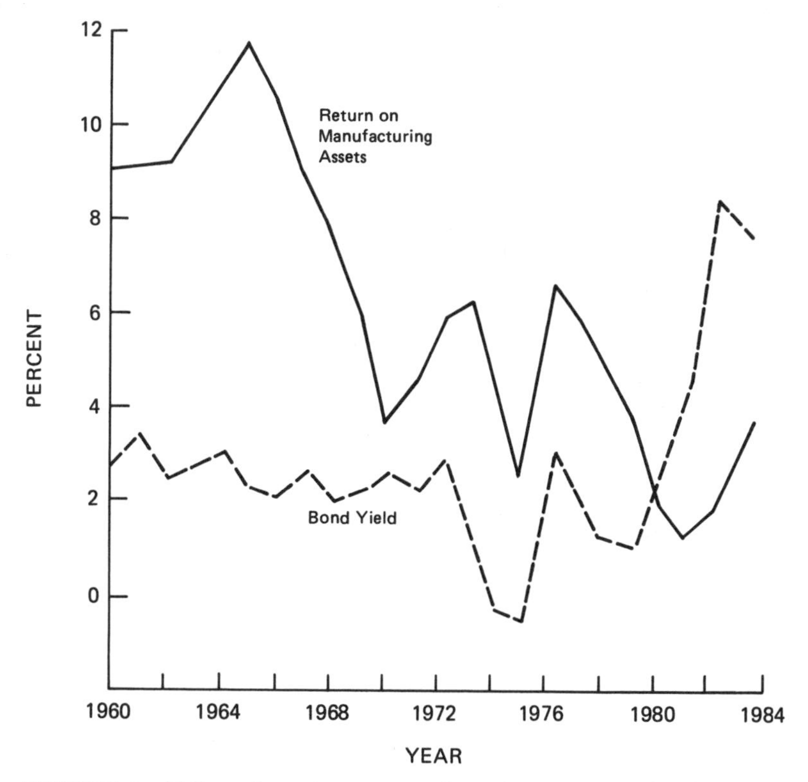

FIGURE 2 U.S. real return on manufacturing assets and industrial bond yield, 1960-1983. Sources: Quarterly Financial Reports of Mining, Manufacturing and Trade Corporations, Federal Trade Commission, 1960-1961; Census Bureau, 1981-1984; inflation data from Economic Report of the President, 1984; Moody's Industrial Bond Yield from Survey of Current Business (July 1984); and Business Statistics, 1979 and 1982.

cent since 1980.[11] Despite this dramatic improvement, estimates of the cost advantage of Japanese producers have grown from $1,000-$2,000 per car in 1979-1980 to $2,000-$2,600 in 1985.[12] Based on consumer surveys, U.S. cars also have lagged behind Japanese makes in quality. U.S. producers have launched new projects—General Motors' Saturn, Chrysler's Liberty, and Ford's Alpha—to eliminate these gaps by rethinking management concepts, employee relations and compensation, and technology. All three companies also are aggressively pursuing joint ventures with foreign producers and captive imports of finished vehicles and parts from several countries to offset the cost disadvantage of domestic production. The companies' approaches differ, but these programs clearly indicate that senior managers in the U.S. automobile industry recognize both the shortcomings of traditional practices and the opportunities that new technologies and new management approaches will provide. Although these efforts may solve current competition problems with Japan, in many cases they will not become operational until about 1990. By that time, other disadvantages and new competitors may have emerged that U.S. firms will be ill-equipped to address.

- Companies in other industries also have aggressively pursued a strong competitive position only to be confronted by intensified competition. Black and Decker Corporation, for example, has devoted significant effort to reducing costs and increasing efficiency by pursuing new investments, increasing automation, reducing its work force, and standardizing parts and product lines across its international operations. Despite these long-term efforts, the company faced growing competition in the world power tool market from Makita Electric Works, Ltd., of Japan and lost a significant part of its market share. Since 1980, Black and Decker has spent $80 million on plant modernization,[13] cut its work force by 40 percent, and adopted new manufacturing practices. The company has regained a 20 percent share of the world market in power tools at the cost of reduced profits resulting from price pressures from the Japanese company.[14] The efforts by Black and Decker indicate the kinds of commitment that are absolutely necessary to maintain a worldwide competitive position.
- The Japanese are not always the prime competitors. Cha-

parral Steel Company, a minimill operation based in Midlothian, Texas, figures that if it can produce steel at a labor cost per ton no higher than the per-ton cost of shipping steel to this country from Korea, it can beat Korean producers. In achieving this goal, the company has invested in some of the most modern steel plants in the world[15] and can produce steel using 1.8 man-hours per ton, compared with 2.3 for the Koreans and 6+ for integrated U.S. producers. Although its capacity and range of product are more focused than that of large integrated producers, Chaparral illustrates two important aspects of the new manufacturing environment. First, the company's experience (and that of other minimill operations, such as Nucor Corporation) has shown that U.S. producers can be world leaders and can pose more of a threat to traditional U.S. manufacturers, albeit in a relatively narrow product line, than foreign competitors. Second, traditional competitive targets, such as matching the production costs of competitors, may not be enough to ensure long-term competitiveness; other targets, such as the shipping costs used by Chaparral, may need to be considered.[16]

- A final example comes from the computer disk drive industry. Floppy disk drives are used in lower-end home computers and personal computers, whereas rigid disk drives are used most often in advanced personal computers and engineering work station products. The disk drive industry was created by U.S. electronics firms from technology developed by International Business Machines and Control Data Corporation. Several smaller firms entered the field in the mid-1970s and quickly grew to substantial size. In recent years, however, the status of U.S. disk drive manufacturers has changed considerably. The leading U.S. maker of floppy disk drives in 1980 was Shugart Associates. Xerox Corporation, the owner of Shugart, has since announced the closing of the unit as a disk drive manufacturer. In 1984, more than 20 Japanese firms manufactured floppy disk drives; no U.S. manufacturers did so. While the United States retains a strong position in rigid disk drives, the Japanese are likely to dominate the next-generation technology, optical disk drives.[17] Developments in this industry show that being the first to market, even with high-technology products, is not a long-term advantage. Constant improvement in both products and processes is needed to ensure survival.

As these examples illustrate, pervasive and potentially damaging change is overtaking U.S. manufacturing across the spectrum of industries from traditional to "high tech." Industries as diverse as motorcycles, consumer electronics, and semiconductor memories also have been subject to lost leadership and declining market shares. Many firms recognize the change and are responding, though often in limited ways. Many more do not recognize the problem or think that it does not apply to them or their industries. Still others attributed their difficulties to the recent high value of the dollar and are looking forward to the benefits of the recent dollar depreciation.

Factors such as interest and exchange rates and unfair foreign competition do have significant effects on industrial health. Unfavorable trends in these areas, however, provide easy scapegoats and disguise other important factors that are changing the manufacturing world. A majority of U.S. manufacturers need to recognize that lowering the cost of the dollar in international currency markets, while important, will not solve all their competitive problems—the price elasticities of many important U.S. imports and exports will determine the long-term effect of the recent decline in the dollar. Although some U.S. commodity exports, such as timber, coal, and some agricultural goods, are likely to increase as the dollar declines, exports of capital goods and major imports of items such as machine tools, automobiles, and consumer electronics may change little as the dollar's value changes, at least in the near term.[18] Many consumers continue to prefer foreign goods because of perceived quality and reliability advantages over their U.S. counterparts. Furthermore, many foreign companies in a range of industries have advantages in production costs that permit them to offset even unexpectedly large devaluations of the dollar by limiting price increases in the U.S. market.[19]

More U.S. firms need to join the minority that recognize the challenges emerging in manufacturing and are devoting resources to meet them.[20] Although competitive challenges are spreading to more and more products and industries, too few companies are making the essential commitment to competitive manufacturing operations in the United States. The rising competition from previously weak or nonexistent sources is prompting a response, but it is insufficient. The initial, and natural, reaction is to do everything better. Redoubled efforts are nearly always beneficial; it is a rare company that does not have room to improve. Doing things better than yesterday or better than competitors today, however, will not necessarily ensure long-term competitiveness.[21]

Another response has been to move production facilities offshore, through foreign direct investment, outsourcing, joint ventures with foreign producers, or other mechanisms. While such arrangements have clear short-term advantages in terms of foreign market penetration and labor cost containment, the long-term repercussions of offshore production strategies are not clear. In some industries, firms must move constantly in search of even lower wage rates; in others, host countries insist on domestic content, technology transfers, and domestic equity positions that lead to independent, competitive production capabilities. Factors vary across industries, and some firms in labor-intensive industries may have no choice but to move production offshore or purchase components or products from abroad. As technological developments yield effective alternatives to offshore production and conditions for foreign direct investment become more stringent, a better understanding is needed about the effects of offshore production strategies on the long-term interests of individual firms and the domestic industrial base.[22]

Another response from U.S. manufacturers has been based on the widely held idea that technology will solve the problem. Advanced manufacturing technology can provide dramatic improvements in efficiency, but only if the groundwork is laid. The benefits of new technology will not be fully achieved if the organizational structure and decision-making process are not changed to take advantage of available system information, if the work force is not prepared for the changes brought by the technology, and if potential bottlenecks created by automating some operations but not others are not foreseen and avoided. Many companies that have powerful computer-aided design (CAD) systems, for example, are using them as little more than electronic drafting boards, negating many of the capabilities that CAD provides because appropriate adjustments in the organization have not been made.[23] Managers need to understand that technology is both a tool for responding to competitive challenges and a factor causing change in manufacturing.

Recent economic data and the experiences of specific industries suggest that a strong case can be made that U.S. manufacturers, with the exception of a handful of enlightened companies, are not responding adequately or entirely appropriately to new competitive challenges, even as those challenges intensify. The first corrective step is to convince managers that they face a manufacturing problem that new technology, offshore production, changes in exchange rates, and redoubled efforts cannot resolve. The next step is to indicate the kinds of changes in manufacturing organizations that will be needed to maintain long-term competitiveness.

The changes needed can be described broadly as a shift from the traditional management goal of maximizing stability, productivity, and return on investment in the short term to the new goal of maximizing adaptability to a rapidly changing market, with long-term competitiveness as the first priority. A number of authors have detailed the changes that are necessary in the management of the manufacturing function.[24] Hayes and Wheelwright, for example, describe the needed changes as a shift from an "externally neutral" role for the manufacturing function, in which the firm only seeks to match the process capabilities of its competitors, to an "externally supportive" role, in which process improvements are continually sought and implemented in an effort to maintain a lead over competitors, and manufacturing is viewed as a significant contributor to the firm's competitive advantage.[25] This shift cannot be made overnight, and it is far too easy to backslide once a new plateau is reached. The shift requires changes in organizational structure and decision-making processes, and it demands new skills: managers must learn to manage change.

STAKES FOR THE U.S. ECONOMY

Because manufacturing remains crucial to national economic and defense interests, the repercussions of declining competitiveness could be devastating. Many economists argue that continued erosion of the domestic industrial base is limited because manufactured goods are the major component in international trade. The United States will remain a major manufacturer because world markets will not tolerate a constant large U.S. trade deficit. Exchange rates will adjust to ensure that the United States can export manufactured goods. Alternatively, the United States will suffer a recession that will dampen demand for imports and alleviate the trade deficit. Recent historical evidence for this argument, however, is ambiguous at best: the United States managed only small surpluses in manufactured goods during the late 1970s, when the dollar was relatively weak, and had a small deficit in the recession year of 1982 (see Table 4). Particularly because exchange rates increasingly react more to financial flows than to goods flows, the sustained process of devaluation of the dollar necessary to maintain the competitiveness of U.S. manufacturers would be difficult to accomplish both economically and politically.

Recessions and shifts in currency value can be painful ways for the nation to reach equilibrium in its manufacturing trade. An alternative is for U.S. manufacturers to implement the organizational, managerial, and technical changes necessary to maintain a strong manufacturing sector. Competitiveness would be based on leadership in product performance and quality rather than on a declining exchange rate. The resulting strength would provide the basis for continued economic growth and provide crucial advantages in areas of national importance, such as

- *Defense*—Counter to conventional ideas that a strong industrial base is necessary to meet U.S. defense commitments, some economists have argued that these commitments can be met without broad support for manufacturing.[26] Although it may be possible to meet them through selective policies designed to support specific defense production instead of entire industries, such an approach would be inadvisable for two major reasons. The first is that it would not provide the productive capacity needed for surges or mobilization in the event of a prolonged conventional engagement. The second reason is that selective policies would hinder the ability of defense contractors to maintain broad technological superiority. This, in turn, would limit their flexibility in response to new defense needs. Production capacity and the technological level of weapons systems are closely linked. Advanced weapons that maintain the qualitative advantage built into the U.S. defense posture require complex manufacturing processes and advanced production equipment, which in turn require broad-based manufacturing capacity.[27] Both the weapons and the production processes are most effectively developed and implemented in the broad context of a healthy manufacturing sector.

- *Living standards*—Although an absolute decline in the manufacturing base might be countered in the short term by growth in the service sector, services are unlikely to be able to absorb a large percentage of unemployed manufacturing workers at their customary level of wages and benefits. Economists disagree about the validity of projections of a shrinking middle class, but declining manufacturing employment would certainly have a large impact on total wage and benefit packages.[28] The increased competition for jobs in services, as well as the likely increase in competition among firms in that sector, would moderate wage growth in services.

International competition in services also can be expected to intensify and moderate wage levels. Apart from the effect of this competition on wages, sufficient growth in services is not at all ensured because many services are tied to manufacturing; if manufacturing decays, these services will decline, too. Furthermore, there is no guarantee that the United States can maintain a comparative advantage and large trade surplus in services that would be necessary to pay for manufactured goods.[29] It is not at all clear that the nation's long-term economic strength lies in services or that potential strength in services is greater than potential strength in manufacturing. Given these considerations, the extent to which services can absorb workers displaced from a declining manufacturing sector and drive overall economic growth remains in doubt.

A technologically advanced manufacturing sector also would result in displacement of workers, but in a competitive, dynamic economy that should be much more successful at creating new jobs. The development of new products, technologies, and support needs would create whole new industries with job opportunities that would be unlikely to develop in a stagnating manufacturing sector.

- *National economic and political goals*—In the domestic economy, regional concentration of manufacturing activity creates the potential for economic disruption from a declining manufacturing sector that would be disproportional to its share of the gross national product. The decline of whole communities dependent on a single factory is, of course, not a new phenomenon, but past experience has clearly shown that the necessary adjustments are difficult and costly. Services in those regions and communities tend to depend on manufacturing and are ill equipped to provide employment and generate income in the face of a declining industrial infrastructure. The decline of U.S. manufacturing would have a severe adverse effect on these regions, and the national policies that would be necessary to support them would be politically difficult to enact. These patterns of regional strength and weakness serve to exacerbate the national economic dilemmas posed by a decline in U.S. manufacturing.

On an international scale, the sheer size of the domestic market is a major driver of economic development, competition, and continued advances on a global scale as a growing number of foreign manufacturers compete for a share of the U.S. market. A declining ability of the United States to supply its own manufactured goods, however, would fundamentally change the relationship between this country and foreign manufacturers. Domestic companies would have less revenue and incentive to pursue strong research and development programs, leading to less innovation and invention and fewer patents. The lack of manufactured goods to trade and of manufacturing income to purchase foreign goods would reduce the bargaining position of U.S. producers and the attractiveness of the U.S. market.

These points illustrate the importance of a strong manufacturing sector and the danger of considering the demise of U.S. manufacturing in purely economic terms. Clearly, each individual industry need not survive, but manufacturing as an economic activity is too important to let decay. Changes in management, process technology, corporate organization, worker training, motivation, and involvement, and government policies are necessary to ensure that resources flow to manufacturing. Changing traditional ideas about education, the role of workers, investment in research, development, and innovation, and overall attitudes toward manufacturing will require input and active participation from a variety of sources. The transition will not be painless. Job displacement, plant closures, and changing industrial patterns will be the norm, as they always have been. But these events will take place in a dynamic economy and, therefore, will be accepted and resolved as smoothly as possible. The result will be a competitive manufacturing sector, far different from today's, that will be a leader in the new era in manufacturing.

This prospect of a dynamic manufacturing sector is, in fact, the Manufacturing Studies Board's vision for U.S. manufacturing. The nation's manufacturers are likely to succeed in making the transition from stable, controlled enterprises to flexible, adaptable businesses that use the latest technology to keep pace with rapidly changing consumer demands and competitors. Timing is essential, however. Market pressures should force these changes eventually, but by then most U.S. firms would be a step behind. The Manufacturing Studies Board hopes to stimulate debate on the nature of the challenge and to hasten the response of U.S. manufacturers, government officials, educators, and labor leaders to the new environment. The following chapters sketch both the technological and human aspects of manufacturing in the future.

NOTES

[1]Time trends for the components of value added in manufacturing are clearly illustrated by the following graph. See Miller, Jeffrey G., and Thomas E. Vollman, 1985, The Hidden Factory, Harvard Business Review 63(5):142-150.

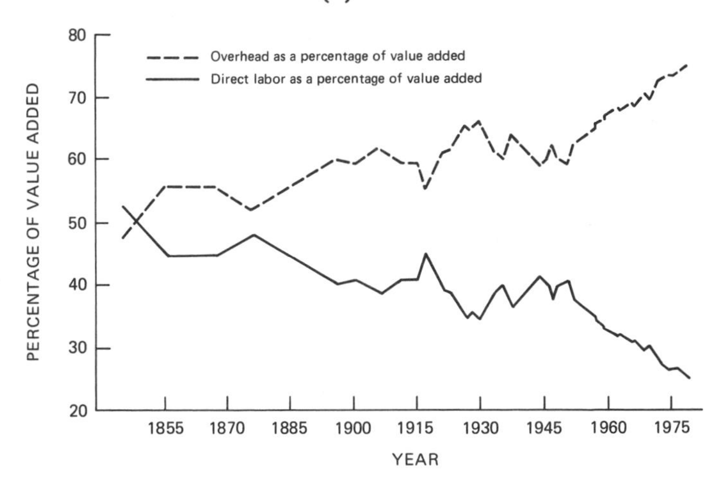

[2]A number of papers addressing aspects of the evolution of the manufacturing environment and manufacturing management from a historical perspective can be found in Clark, Kim B., Robert H. Hayes, and Christopher Lorenz, eds., 1985, The Uneasy Alliance: Managing the Productivity Dilemma. Boston, Mass.: Harvard Business School Press.

[3]Much of the significant drop in the growth of manufacturing output since the 1970s can be attributed to short-term, transient factors, making long-term projections of output growth difficult at best. Long-term manufacturing output as a percentage of U.S. GNP has remained fairly stable at around 22-24 percent.

[4]In particular, see Skinner, Wickham, The Taming of Lions: How Manufacturing Leadership Evolved 1780-1984, pp. 63-110 in The Uneasy Alliance.

[5]Wheelwright, Steven C., and Robert H. Hayes. 1985. Competing Through Manufacturing. Harvard Business Review (January/February), pp. 99-109.

[6]Olmer, Lionel. 1985. U.S. Manufacturing at a Crossroads:

Surviving and Prospering in a More Competitive Global Economy, p. 4. Washington, D.C.: U.S. Department of Commerce, International Trade Administration.

[7]For a detailed analysis of the composition of U.S. manufacturing investment, see Eckstein, Otto, Christopher Caton, Roger Brinner, and Peter Duprey, 1984, The DRI Report on U.S. Manufacturing Industries, pp. 21- 28, New York: McGraw-Hill Book Company.

[8]The issue of high capital costs is given particular emphasis in The President's Commission on Industrial Competitiveness, 1985, Global Competition: The New Reality, Vol. 2, p. 113, Washington, D.C.: U.S. Government Printing Office. Also see pp. 28-35 in The DRI Report on U.S. Manufacturing Industries.

[9]For a more elaborate discussion of the factors influencing managerial investment decisions, see Gold, Bela, 1979, Productivity, Technology and Capital: Economic Analysis, Managerial Strategies, and Governmental Policies. Lexington, Mass.: D.C. Heath-Lexington Books.

[10]For example, see Lawrence, Robert Z., 1985, Can America Compete? Washington, D.C.: The Brookings Institution.

[11]Business Week. April 22, 1985. Can Detroit Cope This Time?, p. 84.

[12]U.S. Congress, Office of Technology Assessment. 1986. Technology and Structural Unemployment: Reemploying Displaced Adults, p. 388. Washington, D.C.: U.S. Government Printing Office.

[13]This figure represents approximately 5 percent of 1984 sales of $1.7 billion.

[14]Business Week. August 26, 1985. Fighting Back: It Can Work, p. 64.

[15]The company uses electric furnaces and scrap iron to pour slightly more than 1 million tons of steel per year, compared with an average capacity for an integrated steel plant of 3.2 million tons per year.

[16]Fighting Back: It Can Work, p. 64.

[17]Note on Small Companies in the Disk Drive Industry. 1985. Palo Alto, Calif.: Stanford University Press.

[18]Porter, Michael. 1986. Why U.S. Business is Falling Behind.

Fortune 113(April):255-262.

[19]For an analysis of the relationship between a dollar devaluation and the trade deficit, see Karczmar, Mieczyslaw, 1985, A Weaker Dollar Won't Slow Imports, The New York Times, October 20, 1985, p. F3. In its *1986 World Economic Outlook*, the International Monetary Fund estimates that if the dollar remained at its March 1986 level, the United States would still have a current account deficit of $100 billion per year. As a specific example, between September 1985 and March 1986, the dollar depreciated by about 27 percent against the Japanese yen, yet Toyota raised prices in the U.S. market by an average of only 3 percent in January and by an additional 4.3 percent in March. See The Washington Post, January 7, 1986, Toyota Raises Prices on U.S. Exports by 3%, p. D2; and Seaberry, Jane, Dollar Drops to Post-War Low Against Yen, The Washington Post, March 18, 1986, p. E1.

[20]For example, 48 percent of U.S. manufacturing executives responding to a Business Week/Harris poll think their own companies have done a "pretty good" job fighting off Japanese competition. See Fighting Back: It Can Work, p. 68.

[21]These concerns are reflected in a report of the Manufacturing Futures Survey Project, initiated at Boston University in 1981. In an effort to understand the manufacturing strategies of companies in Japan, Europe, and North America, a survey of nearly 1,000 manufacturers was conducted over two years. Although the results are subject to various interpretations, a general image that emerged is that American (and European) manufacturers are focusing on improved quality and delivery times, while the Japanese are striving for better cost and flexibility because they have already met the quality and delivery demands of the market. Consequently, American firms seem intent on correcting past deficiencies—doing the traditional things better—and Japanese firms are trying to strengthen and develop an additional competitive edge. Even if they achieve their goals, American firms run the risk of facing a new threat in the future with continued erosion of competitiveness. See Ferdows, Kasra, Jeffrey G. Miller, Jinchiro Nakane, and Thomas E. Vollmann, 1985, Evolving Manufacturing Strategies in Europe, Japan, and North America, Boston, Mass.: Boston University.

[22]The Manufacturing Studies Board is planning a major study of the reasons U.S. manufacturers move production offshore and the repercussions of this strategy for the U.S. economy.

[23]McKinsey and Company. 1983. Computer Integrated Manufacturing. Cleveland, Ohio: McKinsey and Company.

[24]Recent books on manufacturing management include Abernathy, W. J., Kim B. Clark, and A. M. Kantrow, 1983, Industrial Renaissance, New York: Basic Books; Goldrath, E. M., and J. Cox, 1984, The Goal—Excellence in Manufacturing, Croton-on-Hudson, N.Y.: North River Press; Schonberger, Richard J., 1982, Japanese Manufacturing Techniques, New York: Free Press; and Skinner, Wickham, 1985, Manufacturing—The Formidable Competitive Weapon, New York: John Wiley and Sons.

[25]Hayes, Robert H., and Steven C. Wheelwright. 1984. Restoring Our Competitive Edge: Competing Through Manufacturing. New York: John Wiley and Sons.

[26]For example, see Can America Compete?

[27]The argument that sophisticated weapons systems require advanced manufacturing technology is made more fully in Committee on the Role of the Manufacturing Technology Program in the Defense Industrial Base, 1986, The Role of the Department of Defense in Supporting Manufacturing Technology Development, Washington, D.C.: National Academy Press.

[28]Both sides of the argument are presented in Work in America Institute, 1985, Middle Class Shrinking as Pay Inequality Grows, World of Work Report (August), p. 5.

[29]For example, the New Development Corporation (NDC) in Taiwan provides engineering services for the development of computer hardware and software for International Business Machines Corporation (IBM). NDC engineers, with reportedly higher productivity, are paid one-fourth the salary of American engineers. Subcontracting such services was an unprecedented step for IBM but has proven so successful that other American electronic and computer firms are considering establishing their own engineering and development operations in Taiwan. See Stokes, Bruce, 1985, Rising Trade Deficit, High-Tech Growth Are Threats to U.S.-Taiwan Relations, National Journal (November 30), pp. 2696-2702.

2
The Role of Advanced Technology in Future Manufacturing

The recent performance of U.S. manufacturing (see Chapter 1) suggests that relatively few firms have begun to make the changes necessary to compete in the manufacturing environment of the future. As competition intensifies and technological change affects both products and processes, standard practice will become less and less effective. In many cases, success in the marketplace will require reassessment of total business strategy and the tools used to pursue it. Functions such as purchasing, marketing, and distribution will need to adapt, but the primary need in many industries will be suitable design and production strategies. In fact, the effectiveness of the design and production functions in supporting overall business strategy may prove to be the major determinant of the competitiveness of U.S. industry.

Effectiveness is a relative term that should not necessarily be equated with the most advanced design and production technologies. Selecting the proper technologies is a major management challenge that will become more complex as technology advances. The selection will depend on factors such as the capabilities of the technology, the type and number of products being manufactured, the relative costs of production, the abilities of competitors, and the long-term objectives of the company. Advances in flexible manufacturing systems (FMS), for example, will be more valuable to a batch manufacturer with frequent modifications in machining requirements than to a mass producer of standard parts with infrequent design changes. Each manufacturer, even each plant, must assess its objectives and employ technology accordingly.

Management also must implement technology effectively. Effective implementation requires recognition of the effects that new technologies can have on total operations. Material handling systems and automated assembly systems, for instance, may be installed to reduce direct labor inputs, but they also can result in lower inventory costs, a streamlined production process, and new part designs. Such systems, however, may create a need for additional computer hardware and programming talent as well as specialized engineers, so total labor costs may not change much. Managers also must understand that needs and capabilities may change rapidly. None of these technologies can be once-for-all investments. The best approach may be a modular system based on a long-term strategy of integrating "islands of automation" into a single system with a comprehensive data base and communication network.

Appendix A provides a description of the major technologies that are expected to be available to manufacturers in the future. This chapter, instead of focusing strictly on the physical capabilities of the technology, emphasizes issues in using advanced technologies in design and production processes—issues that may represent radical departures from traditional manufacturing experience. Some companies are already confronting these issues in their efforts to modernize production facilities. These U.S. manufacturers are well into the early stages of a long-term technological revolution in manufacturing. Their early experiences give strong indications of the direction that manufacturing management, strategy, and competition will take in the future.

New technologies will permit manufacturers to implement strategies that previously were, at best, ideal objectives subject to a range of compromises. These strategies will vary by company, but several objectives will be common to all. Every manufacturer will continue to strive for more efficient use of total resources. Under this overriding goal, key objectives can be summarized as (1) more rapid response both to market changes and to changing consumer demands in terms of product features, availability, quality, and price; (2) improved flexibility and adaptability; and (3) low costs and high quality in design and production. More specific categories could be added, but these three broad objectives are likely to be major elements of successful competitive strategies. Clearly, new technologies alone cannot achieve these goals. Great strides can be made using new management approaches, more flexible work rules, and organizational changes. In fact, these factors (see Chapter 3) are absolutely crucial to improving manufacturing effectiveness, particularly since new technologies are not likely to achieve optimal results without modifications in the human resource aspects of the organization. Managerial and organizational changes are necessary conditions for improved manufacturing competitiveness—new technologies can increase those benefits exponentially.

RESPONSIVENESS

A manufacturer's response to change varies with specific circumstances, such as the structure of the industry, the particular product line, the market in which the product is sold, and the firm's competitive emphasis. Because of the many permutations of external factors and possible responses, the important requirement is not a strategy for every contingency but the ability to pursue a range of strategies aimed at particular combinations of circumstances. Much improvement in responsiveness can be achieved by reevaluating the company's operations, particularly in design, engineering, and manufacturing, to determine handicaps, improve functional cooperation, and strengthen common goals. Cooperation and even integration of the many functions in the entire manufacturing system will need to be pursued aggressively to achieve many of the necessary improvements in producibility, productivity, quality, and responsiveness. In the future, these efforts will be strengthened by the capabilities embodied in advanced manufacturing technologies. When combined with effective organizational changes, advanced manufacturing technologies will be powerful tools for achieving enhanced responsiveness to many external factors that affect design and production.

The technologies that can help maximize responsiveness will vary among firms and product types, and an enormous amount of tailoring will be involved. A major consideration will be the degree to which the firm's competitive strategy depends on price leadership or product differentiation. A low-price strategy implies the ability to offer the combination of performance and quality demanded by customers at the lowest possible price, yet still respond to demand changes, variations in input availability and relative costs, and changes in competitors' capabilities. A product differentiation strategy, on the other hand, aims to supply a range of price and performance options that covers most consumer demands. A firm with a low-price strategy may benefit more from improved production, material handling, and inspection technologies, while a product differentiation strategy may demand more emphasis on computer-aided design, flexible manufacturing systems, and new materials and processing techniques.

For many manufacturers, a CAD system is a necessary first step in improving responsiveness. A CAD system that includes a comprehensive CAD data base and modeling capabilities can provide significant benefits beyond such simple use as an electronic drafting board. It can speed changes in product design to meet market demand by improving the productivity of design, manufacturing, and process engineers and ensuring that new designs are readily producible. When coupled with an effective group technology (GT) data base (see Appendix A), such a system permits parts to be designed and produced rapidly and effectively with minimum changes in tooling, numerical control programs, and material requirements because new parts will resemble current parts as much as possible. Of course, designing for producibility is a crucial goal regardless of the technology used, but a CAD system can improve the designer's ability to simplify new and existing part designs, improving manufacturability and reducing the need for complex multiaxis machining.[1] The benefits in terms of design costs, design-to-production times, and quality can far outweigh investment costs. The full benefits will not be achieved, however, without vigorous efforts by management to train engineers and technicians, encourage acceptance, and change the corporate culture, support systems, and business philosophy in a way that will allow the capabilities of the CAD system to permeate the company. Improved design response does no good if new designs are allowed to languish on the factory floor.

Although some type of CAD system and GT data base will be a major factor in most firms' efforts to maintain competitive response times, the types of process technology that offer substantial benefits will vary tremendously among firms depending on the types, variety, and volume of parts produced. With batch parts, for example, numerically controlled (NC) machining centers and turning centers with cutting tool drift sensors, self-correcting mechanisms, and laser-based inspection equipment should provide significantly better quality, reduced scrap and rework, and shorter response times. The manufacturer could be more confident that each part was within specifications and, therefore, could respond more quickly and effectively to changing consumer demands.

A flexible manufacturing system (FMS) would take these benefits a step further. An FMS can perform a number of operations on a limited variety of parts within a family (e.g., parts having a fundamental aspect in common) without the need for manual handling of the parts between operations. Given this definition, a single complex NC machining center conceivably could be described as an FMS. More sophisticated FMSs might combine a number of machining centers with material handling systems and automatic tool and part changers and integrate the required CAD, GT, material, tooling, and NC program data. In the future, automated assembly operations will be included in the FMS installation, eliminating much of the required human intervention, but this development depends on progress in a number of technological areas. (Appendix A includes a detailed description of FMS technology.) Current FMS installations cover a range of equipment combinations, complexity, and capabilities, which indicates that manufacturers' production operations must be evaluated carefully in developing an FMS that is effective and appropriate for production needs. For many applications, responsiveness through rapid set-up and delivery is as beneficial as the flexibility offered by an FMS. In many other cases, however, FMSs provide more complexity, capabilities, and costs than are really necessary for the firm's production, and alternate approaches, such as design simplification or hard automation, would be more appropriate.

Although most mass production industries may not need the capabilities promised by an FMS, computer-integrated manufacturing (CIM) systems will find applications in all types of manufacturing regardless of the process machinery on the factory floor. CIM refers basically to the data-handling capabilities of the manufacturer. It is a sophisticated system for gathering, tracking, processing, and routing information that links purchasing, distribution, marketing, and financial data with design, engineering, and manufacturing data to expand and speed the knowledge available to employees and managers. CIM systems will use interactive data bases and hierarchical control systems coupled to advanced CAD systems, modeling and simulation systems, computer-aided engineering (CAE) systems, production process planning systems, flexible manufacturing systems and/or hard automation processes, material handling systems, and automated inspection/quality assurance systems. (Appendix A provides more technical details of each of these systems.) At a minimum, each of these elements will be necessary for an effective CIM system because each builds and depends on the others. Each also will require substantial investment in hardware, software, and personnel, but the capabilities provided by CIM can be expected to introduce new priorities and new determinants of competitive advantage.

In essence, a complete CIM system will allow teams of design, process, and manufacturing engineers to design products quickly in response to market demand, product innovations, or input price changes. The CAD system, modeling system, CAE systems, GT data base, and other manufacturing data bases will provide feedback on producibility, material requirements and availability, various production process options, costs, and delivery schedules. The design data will then be transferred directly to the other elements of the CIM system, and the new product can be produced within a very short time. For many products, the interaction of the engineers with the CIM system may be the only significant direct human participation in the production process. Such a capability has already been demonstrated, at least for simple metal parts in very low volume, at the National Bureau of Standards' Automated Manufacturing Research Facility, as well as by several private manufacturers. Substantial research will be required, however, to expand and enhance both the hardware and software technology to allow high-volume production of complex parts (see Appendix A).

Manufacturers in the future are likely to find that the responsiveness provided by CIM is a necessary competitive advantage. By providing virtually immediate information on every aspect of the manufacturing enterprise, CIM will force manufacturers to eliminate delays in design and production, because rapid reaction to new opportunities and changing conditions will be a major competitive factor. These developments will introduce new bases for competition and may change the economic environment in many industries. Competition will be based on management and labor skill, organizational effectiveness, the price and quality of the final product, speed of delivery, and serviceability of the design, including appropriateness of materials used, functionality, ease of repair, and longevity. Since CIM technology will be readily available, market success will depend on proprietary refinements to the CIM system and how well it is used. The manufacturer who can use CIM not only to respond quickly but also to minimize total resource requirements will have a competitive edge.

FLEXIBILITY

Responsiveness, as discussed here, refers to a manufacturer's ability to react quickly to changes in external conditions; flexibility is really an extension of that concept to internal factors. In fact, one can distinguish several types of flexibility.

- Process flexibility is the ability to adapt processes to produce different products without major investments in machines or tooling for each product. This type of flexibility is the cornerstone of flexible automation technologies that allow optimal matching of materials to product applications, the ability to use various materials, and the ability to produce a variety of product designs.
- Program flexibility allows process path modifications, adaptive control and self-correction, unattended operations, and back-up capabilities to maintain production even when some part

of the process fails. This type of flexibility addresses the need to optimize equipment use and run multiple workshifts.

- Price-volume flexibility provides the ability to maintain economic production in a wide range of market conditions resulting from cyclical and seasonal changes in demand. (This type of flexibility concerns external circumstances, but the firm's internal flexibility will determine the success of the response.)
- Innovation flexibility refers to the ability to implement new technologies as they become available. This type of flexibility depends on a modular approach to manufacturing systems integration that is essential to an evolving design and process capability.

Manufacturers have always confronted the problems these various types of flexibility address and to solve them have relied on compromises between production capabilities and costs. The costs of increased flexibility traditionally have included higher inventory, increased tooling and fixturing requirements, lower machine utilization, and increased labor costs. By reducing some of these costs through advanced manufacturing technologies, more producible part designs, streamlined organization, better functional cooperation, just-in-time inventory control systems, material requirement planning, and other mechanisms, future manufacturers have the opportunity to reduce the need for many of the traditional cost-flexibility compromises. Cost-flexibility trade-offs will always exist, however, and the advantages of specific technologies will vary tremendously among industries and product lines.

For many mass production industries, conventional hard automation will remain the most efficient production process. Hard automation—that is, transfer machining lines—is relatively inflexible; it is generally product specific. Advances in design capabilities, sensors, materials, robotics, material handling systems, and automated inspection technologies should introduce a degree of flexibility into these operations, but efficient production will still tend to depend on economies of scale and product standardization. For traditional batch part manufacturers, the flexible automation technologies embodied in an FMS will change many of the historic cost-flexibility compromises. In some applications, FMSs with flexible fixtures can be expected to reduce set-up times to near zero, allowing smaller and smaller lots produced on demand to become both economically feasible and competitively necessary. In other applications, group technology, designing for producibility, and efforts to speed changeovers will increase flexibility at less cost and with greater effectiveness than elaborate FMS installations.

Although the flexibility of the process equipment on the factory floor will differ between mass producers and batch manufacturers, both will benefit from the flexibility embodied in CIM. The ability to gather and manipulate data in real time as orders are received and products are made will provide a degree of control over the manufacturing process that has not been possible in the past. It is in this context that the issues and benefits of flexibility become particularly relevant to a competitive production strategy.

No manufacturer could afford and no technology could provide infinite flexibility, and increasing investments in advanced technologies will not necessarily correlate with increasing flexibility in production. As an extreme example, a machine shop using manual machine tools and expert craftsmen may be more flexible in producing a broad range of parts and may be better able to improvise to produce prototypes than a more modern machine shop using NC machine tools. The manual shop, however, is likely to be less cost effective and slower, have more scrap and rework, and, most importantly, be ill prepared to take advantage of other computer-based technologies that could improve control over and the effectiveness of the total production process. The NC shop may have a narrower product line, but it is likely to have quicker response times, more consistent tolerances, better repeatability, and greater ability to integrate other computer-based technologies, such as CAD, and use them effectively. Strictly from a production perspective, the manual shop could be described as more flexible, but from a total operations perspective, the NC shop is more flexible. Its potential for introducing new design and production technologies, particularly the data-handling capabilities of CIM, is incomparably greater than that of the manual shop. Neither approach is indisputably correct, however; each is based on value judgments and trade-offs made by the owners in response to their circumstances. This example, although extreme, illustrates the unavoidable cost-capability compromises that will always be confronted in the pursuit of greater production flexibility.

Determination of an optimal level of flexibility for a given plant must include not only the cost effects of different ranges of product mix and quality relative to production capacity, but also the cost effects of fluctuations in demand. The greater the investment in production facilities and the associated fixed costs, the higher the break-even rate of capacity utilization. Consequently, cost savings expected from more capital-intensive production systems at high levels of utilization must be balanced against higher average unit costs as seasonal and cyclical fluctuations reduce average capacity utilization. This dilemma involves price-volume flexibility, but the lesson is applicable to all four types of flexibility. Significant planning is required to achieve optimum flexibility; ad hoc programs and investments will be counterproductive.

Although cost-flexibility compromises will continue to apply, the basic flexibility provided by new technology will be much greater than with current NC machining and turning centers or FMS installations. Future flexible manufacturing systems are expected to accommodate a variety of process plans resulting in optimum use of equipment. The system will permit variation in the sequence of operations on the same part, constrained only by the need to drill a hole, for example, before reaming it. This multiple-path flexibility will be feasible not only because set-ups will be flexible and essentially cost-free, but also because part programs will not be specific to any one machine. Machine tools will have sufficient embedded "intelligence" to determine their own parameters for a given part. CIM will ensure that the requisite information is transferred to individual machines from the original CAD data for that part.

This scenario of an FMS and CIM system producing parts in small batches along an optimal processing path (but using multiple processing paths when necessary) from data generated by the CAD system is the basis of flexible automation. The operation of the system will be predicated on additional attributes. To illustrate, the feedback provided by the CAD system to improve the producibility of a given design depends on the production capabilities of the machine tools in the system. Although the system will automatically maintain data on current production capabilities, compromises will undoubtedly arise between maximizing the producibility of a design with current process capabilities and maximizing the functionality of the part or increasing producibility with alternate production processes. The manufacturer may need to compromise with the purchaser to ease the specifications for producibility or increase the price to compensate for a less producible design; subcontract production of the part to a firm with the necessary production capabilities; or add new capabilities to his own production facilities. Even a sophisticated FMS will function effectively only within relatively narrow ranges of material inputs, processing capabilities, and product dimensions. Its potential cost advantages are likely to be threatened when these determinants of production requirements are still subject to substantial change.

The process flexibility discussed at the outset involves these considerations, along with factors such as maintaining optimum equipment use rates while maximizing the overall flow of parts on the factory floor and minimizing routing costs. This type of flexibility also would include the ability to accept data from subcontractors' CIM systems, determine the most efficient producer of the part, and integrate that subcontractor's production and schedule with the other aspects of the total order. If the part is not subcontracted, the CIM system will need to be sufficiently flexible to allow rapid acceptance and integration of new process capabilities. Continued development of new technologies is likely to increase the importance of innovation flexibility; to maintain competitiveness, the CIM system will need to be modular, using computer architectures and interfaces that permit additions to the system to be made easily (see Appendix A).

The system should not be so integrated, however, that a single failure can stop the entire factory. Program flexibility should provide every CIM system with sufficient back-up to maintain production even if relatively crucial parts of the system fail. The risk of failure of each component of the system will be weighed against the cost of providing back-up for the component or some other acceptable alternative to maintain production. A wide variety of solutions to this difficult dilemma can be expected.

Effective implementation of these technologies will require adjustments from managers, engineers, and customers. The tradeoffs between cost and flexibility will vary among industries and products. Advanced CIM systems will not be infinitely flexible because flexibility will depend on management practices and organizational effectiveness, as well as software, tooling, and material availability. In general, however, both mass production and batch manufacturing industries that can take advantage of CIM technologies can expect a degree of flexibility unknown in the past, with benefits in responsiveness, competitiveness, and total production costs that outweigh the cost of the technology itself.

COST AND QUALITY

Advanced manufacturing technologies will give managers new tools to help them minimize use of total resources and thereby reduce product life cycle costs. Whether competitive strategy emphasizes low price or product differentiation, price competition in the future is likely to be severe. Reducing life cycle costs and maximizing quality for every product line will be an important determinant of competitiveness and profitability. Cost minimization must not be pursued, however, at the expense of responsiveness and flexibility, as many manufacturers may be tempted to do. The best way to avoid overemphasis on costs is to think in terms of minimizing use of total resources, not only in production but also in purchasing, design, distribution, finance, marketing, and service. Nevertheless, the attention given to individual production factors will continue to depend to a great extent on relative factor costs and the shifting importance of factors in particular industries.

Because new manufacturing technologies will be developed and implemented at various rates, the effect of technology on relative factor costs is difficult to predict. For some manufacturers, new technology is likely to have only a limited effect on direct labor costs and, indeed, will be applied for reasons other than labor savings. The use of CAD and CIM systems will allow these manufacturers to continue to design and produce parts in the United States, but other operations, especially some assembly operations in which labor remains a high proportion of cost, may be candidates for subcontracting or movement to low-wage countries.[2] The potential savings from low labor rates abroad would need to be balanced against the costs of coordinating demand, production, and delivery. Timely production and delivery will be important in avoiding loss of orders and inventory costs that may not be faced by competitors. These factors will require manufacturers with offshore facilities to use significant forward planning to align production with demand. Advanced technologies will allow manufacturers to handle data in ways that should help to ameliorate the disadvantages of offshore operations, but these gains may not be sufficient to offset the transportation costs, delays, and relative isolation entailed by distant production facilities.[3]

For many manufacturers, advanced technology can be expected to allow more rapid reduction of direct labor inputs, although again, labor savings may not be the major motivation for the investments. A CIM system with flexible automation will permit managers to refine and adjust operations to reflect changes in relative costs over time. It also will introduce entirely new elements to the manufacturer's cost structure, alter traditional ways of measuring costs, and eliminate some major portions of traditional factory costs. It will be possible, for example, to reduce direct labor to insignificant levels or eliminate it in some applications. With no direct labor inputs, some measurements of labor productivity and cost allocation based on direct labor will be irrelevant. New cost accounting systems will be a major need (see Appendix B).

Elimination of direct labor is not the same as eliminating labor costs. Technicians, engineers, and programmers will be needed in increasing numbers to maintain and implement CIM systems. Salaries for these employees are likely to exceed wages for direct labor, and their productivity may be more difficult to measure. Even with higher individual salaries, however, labor costs should decline as a share of total production costs because of the capital investments required to keep a CIM system up-to-date.

It is difficult to predict the effects of investments in new technology on capital costs as a proportion of total costs. Firms will need to monitor the production capabilities of their competitors, as well as those demanded by the market; timely updates of the design and production system will be a competitive necessity. Greater return can be expected and less total capacity may be needed, however, because CIM is expected to allow more workshifts and more optimal use of productive equipment through flexible process plans, less scrap and rework, higher-quality production, and lower product life cycle costs. Justification and amortization of technology purchases will be based on total system performance, which implies a significant shift in the measurement and allocation of capital costs.

Although total capital costs in the future are unpredictable, one element of capital costs—tooling costs—can be expected to increase dramatically over historic levels. Continued implementation of CIM systems and FMSs, along with the emergence of new materials and processes, will greatly increase the volume, variety, and complexity of tooling requirements and the need to move the tooling around the factory. Increased tooling and handling costs are already evident in many manufacturing operations, and the trend can be expected to accelerate as other new developments are implemented.

Developments in new materials and material processing will have a significant impact on material costs and availability, especially vis-a-vis product performance and quality. Advanced material handling systems should have a major effect on the costs of moving and storing materials. New materials such as high-strength resins, composites, and ceramics will create new options in product development, providing significant improvements in performance while reducing material requirements (see Appendix A). Ceramic engine parts, for example, are under development by virtually all major combustion engine manufacturers and will allow simplified engine design and fewer total parts. Once the material and processing problems are overcome, the effects on material costs and requirements will be substantial. Similar effects can be expected with other materials and applications. Even with more traditional materials (e.g., metals), progress in ultraprecision machining will reduce material requirements and improve product performance.[4] These developments will greatly expand the choices available to managers in material application and processing, which will make the data management abilities of CIM virtually indispensable.

For all manufacturers, the ability to accumulate, store, and process data will become a growing force in production, motivated by a rapid decline in the cost of data management. Data will be gathered and accessed rapidly and easily, with a number of important repercussions, one of them being the increasing significance of time as a factor of production. The time between design and production and from order to delivery will shrink dramatically. The trend toward shorter product life cycles and rapid technological developments can be expected to make very small increments—hours and days instead of weeks, months, or even years—crucial factors for competitive production.

The increased ability to manipulate and accumulate all types of data will have a significant impact on plant location decisions. Since information for the entire organization will be available almost concurrently regardless of plant site, the criteria for plant locations will emphasize costs, available process technology, responsiveness, quality, and optimal resource use. Some of the considerations involved in decisions to move plants offshore have been discussed; these considerations may lead to more decisions to keep manufacturing capacity onshore.

In fact, there is some evidence that, due to responsiveness, flexibility, and quality concerns, future trends in factory locations, particularly for component manufacturers, will be toward a proliferation of smaller factories closer to final markets and greater use of contiguous manufacturing, in which progressive manufacturing operations are located in close proximity to each other. New technologies will make both of these strategies easier to pursue for many industries, and market demands may make them a necessity. For some industries, the concept of the microfactory will become important: small factories, highly automated and with a specialized, narrow product focus, would be built near major markets for quick response to changing demand. Because of the unique circumstances of each industry, in terms of technology availability, labor requirements, cost structures, and competitive circumstances, it is difficult to predict how strong each of these trends will be, but they are representative of new options available to manufacturers in their efforts to maximize competitiveness.

In addition to the costs of labor, capital, materials, and data management, CIM can be expected to change other traditional costs in manufacturing. The producibility feedback and modeling capabilities of the CAD system will reduce product development costs, which will allow firms to either reduce prices or do more product development. The monitoring and self-correcting mechanisms embedded in the machine tools will reduce scrap and rework costs. These capabilities also will result in higher quality, which will greatly reduce service costs and attract a broader customer base. Inventory costs also will decline—because set-up times and costs will not be major considerations in the flexible CIM system, it should be possible to make very small batches of products to order without large inventories of materials or finished parts. Finally, the flexibility of CIM will permit companies to use materials with a variety of specifications, customize products, and focus on markets for products with higher value added. The extent to which companies pursue these capabilities will depend on their overall competitive strategies.

All of these considerations imply that manufacturers will have a very different cost structure in the future than they have today. Continuous investment will be a competitive necessity and will be justified on a systems basis. The CIM system will manage the use of material, time, and equipment to such an extent that total inputs and, therefore, total production costs, can be optimized within the limits imposed by the hardware and software capabilities of the moment. Close monitoring of input requirements will be a crucial ingredient, along with responsiveness and flexibility, in determining competitiveness. High quality will be a necessity because customers will expect it; any perceived slippage in quality will cost customer loyalty and market share and obviate many of the benefits of CIM and new materials.

CONCLUSION

The most important factor in improving responsiveness, flexibility, costs, and quality will be the effectiveness of management practices, organizational design, and decision-making criteria. As the capabilities and advantages of new manufacturing technologies progress, they will become increasingly important to managers' future strategies for improving competitiveness. Furthermore, the effectiveness of the technology in accomplishing corporate goals is likely to depend on having the most appropriate hardware and software in the CIM system. Changes and adjustments to the system will be based on each company's market situation, product line, and customer base, so many of these capabilities will be internally developed and proprietary. Along with the management aspects of the manufacturing organization, they will determine competitive advantage in the manufacturing environment of the future.

This view of manufacturing technology is very different from the traditional technical view. Advanced manufacturing technologies are not going to solve all the problems of production. Instead, they will give managers many more options. Managers will have an even greater need to focus the goals of the firm and then assess the needs of the manufacturing function and how technology can best address them. Once choices are made, managers will not have the luxury of running the technology for long periods while they focus on product design, marketing, or some other function to maintain a competitive position. Dynamic, continuous improvement of manufacturing capabilities will become essential to long-term success.

NOTES

[1]Deere and Co. has had much success in simplifying part designs, to the extent that the company is deemphasizing the use of multiaxis machining centers in FMSs because much of the complexity in manufacturing components has been eliminated through simpler designs. See Giesen, Lauri, 1986, Deere Abandoning Focus on Flexible Manufacturing Systems, American Metal Market/Metalworking News (March 24):1,32.

[2]Any of these three operations—design, parts production, and assembly—may be subcontracted in the future to maximize efficient use of available technology.

[3]In many industries, there are already indications that large multinational corporations are becoming disenchanted with a low-wage strategy. The need to move facilities continually as wages inevitably rise in developing countries, the increased viability of automating domestic facilities as an alternative to siting plants in low-wage countries, and the pursuit of long-term production strategies have underscored the costs of a low-wage strategy and other developments have undermined the benefits. See Ohmae, Kenichi, 1985, Triad Power: The Coming Shape of Global Competition, New York: The Free Press.

[4]McClure, E. Raymond. 1985. Ultraprecision Machining and the Niche of Accuracy. CIM (September/October):16-20.

3
People and Organization

The competitive pressures and technological capabilities discussed in the previous chapters are two dimensions of the changes that can be expected in the future manufacturing environment. This chapter addresses changes in the management of people and organizational design that future manufacturers will need to pursue to be successful. Such changes can strengthen the competitiveness of many companies regardless of the technology employed, and in virtually every case, modifications in both the internal and external relationships of the business are a prerequisite to effective use of new technology.

The changes needed in people and organizations will be a difficult aspect of the revolution in manufacturing. They require a dramatic refocus of the traditional culture in the factory, away from hierarchical, adversarial relations and toward cooperative sharing of responsibilities. With such fundamental changes, progress will be slow, the degree of change will vary among companies, and the full transition is likely to be accomplished by a relatively small number of companies. However, the demands placed on manufacturers to be effective in an increasingly competitive marketplace can be expected to push managers and workers in the directions described in this chapter.

Much depends on the size and culture of the firm and the commitment of managers and workers. Many manufacturing enterprises need changes not only in broad organizational areas and management philosophy but also in employee behavior, union policies, and customer-supplier relations. Every stakeholder—managers, employees, owners, suppliers, and customers—must recognize the challenge and be prepared to change traditional practices. Furthermore, people who may not have a direct stake in manufacturing—government officials, educators, researchers and scientists, and the general public—will need to understand the importance of manufacturing to future prosperity, recognize the evolving role of manufacturing in the U.S. economy, and support the many social and cultural changes that will both result from and encourage continued progress in U.S. manufacturing.

A SYSTEMS APPROACH

Part of the problem of U.S. manufacturing is that the common definition of it has been too narrow. Manufacturing is not limited to the material transformations performed in the factory. It is a system encompassing design, engineering, purchasing, quality control, marketing, and customer service as well as material transformation; the operations of subcontractors and the whims of customers are also important parts of the system. The systems approach is a key principle not only for manufacturing technology, but also for organizational structure, supplier relations, and human resource management. Such a concept has been foreign to most U.S. managers (although embraced by Japanese managers), and the result has been a lack of responsiveness and declining competitiveness in many industries. Managing manufacturing as a unified system will profoundly affect every activity involved; it is the only way to take advantage of the many opportunities in both products and processes that the future will bring.

An aggressive systems approach in a company should eliminate many of the functional distinctions that can introduce inefficiencies into the production process. Instead of the labyrinth of functional departments common in many firms, the operations function is likely to become the focus. Ancillary and supportive functions will be reintegrated into operations. Maintenance and process design, for example, will no longer be distinct entities with separate schedules and staffs; instead, employees in operations will be responsible for maintaining equipment or modifying the process as the need arises. Such reintegration will mean that management structures are likely to be more streamlined and that many job classifications will be eliminated to allow employees to perform multiple tasks. Job design and classification will be based on broad operational functions rather than narrowly defined activities.

Functions such as product design, manufacturing, purchasing, marketing, accounting, and distribution will require close cooperation and tight coordination. Eliminating them as separate departments would be impractical—the various types of expertise will still be needed—but, with increasing computerization and communication capabilities, information on each area will be widely available and close cooperation will be essential. This cooperation often is likely to be accomplished through working groups of people from different departments. They will include permanent groups to ensure long-term integration of ideas and temporary groups designed to address specific projects. Techniques such as comprehensive job rotation may be used to eliminate interdepartmental barriers. The process of integrating the data bases and process technologies in the factory also will help to eliminate artificial barriers between functions, but the major tools for change will be the guidance of senior managers and the initiatives of employees.

In external relations, a systems concept calls for reassessment of the mechanisms used to specify, order, manufacture, and deliver subcontracted parts. Because production by suppliers will be viewed as the initial step in the manufacturing system, major customers will need to take a strong interest in the capabilities of their suppliers and institute programs to raise those capabilities through gentle persuasion, direct assistance, or reselection. As an example of the changes that can be expected, customers' design equipment will be able to communicate directly with suppliers' production equipment. Substantial investments will be made in communication linkages to allow extensive sharing of data on design, production scheduling, material requirements planning, and costs.

These types of arrangements will be essential for flexible management of the manufacturing system, but they imply significant change in supplier relations. More subcontracts will be long term, and the number of captive shops supplying one customer can be expected to increase substantially. The investment in communication links by the customer and the corresponding investment that the customer will expect of the supplier will make long-term contracting desirable for both parties. Since long-term contracts weaken the threat of changing suppliers if standards are not met, a strong commitment to close cooperation will be a necessity. Problems will need to be solved as they arise, just as with in-house production, because the cost of failed relationships will be high. Both parties will lose independence in the subcontracting process, but the advantages of an integrated, highly efficient manufacturing system will outweigh the costs.

Examples of these types of relationships can be seen already in both the automobile and aircraft industries. Long-term subcontracting has been common in both industries for years, but only relatively recently have design data been transferred directly from the CAD system of a major manufacturer to the machine tools of the subcontractor.

PARTICIPATION AND OWNERSHIP

A key step in the evolution of human resource management in manufacturing will be to broaden participation in the company's decision-making process. Employees at all levels should be given an opportunity to contribute ideas, make decisions, and implement them in areas that may affect operations beyond the individual's formal responsibilities. The principle involved is intellectual ownership: if all employees can feel a degree of ownership in decisions that affect them and the company, they are likely to support those decisions more enthusiastically, resulting in a highly motivated work force and a more responsive, effective company. Extensive, even universal, participation in decision making gives all employees a stake in the company, beyond financial considerations, that may be essential for continued competitiveness in a rapidly changing environment.

For most manufacturers, such a decentralized decision-making process will require a major cultural shift and a number of prerequisites to avoid disorder. The most fundamental requirement is a well-understood, common set of goals and a high level of commitment to them from both managers and employees. Beyond that, both management and labor must meet certain responsibilities.

Management cannot expect employees to contribute ideas and participate in decision making without the necessary knowledge and expertise. Vehicles will be needed to facilitate the rapid flow of information within the organization and to ensure that the proper intellectual resources are available at all levels. Close links between upper management and operatives on the factory floor will be required for rapid information exchange and responsiveness. Information must flow both upward and downward in the organization. Employees must understand fully the goals and priorities of the firm to make consistent, effective decisions. Managers need to be assured that the correct decisions are, indeed, being made. This type of cooperative, two-way flow represents a radical shift for many firms that may cause significant cultural disruptions.

Information linkages are currently provided by several layers of middle management that serve primarily as an information conduit. With decision-making responsibility pushed to the lowest possible level, the extra layers of middle management are likely to be both unnecessary and unaffordable. The result in most cases probably will not be mass layoffs of middle managers; instead, the change will become manifest as a gradual blurring of the distinctions between operatives and managers. Middle managers will be reduced in number and merged into new roles that allow direct access between upper management and floor workers. Knowledge requirements, authority, and responsibility will tend to converge, resulting in much flatter organizations. This fundamental change in organizational structure already is happening in a number of companies.[1]

Progress in factory communications technology also can be expected to facilitate information flows and contribute to the elimination of management layers. Wide, if not universal, access to all types of information, from part designs and scheduling to accounting data and marketing plans, will reduce the need for personal exchanges of information. Both upper managers and operatives will have direct access to the information needed to make effective decisions. With the process of flattening the organizational structure and creating direct communication channels between operatives and upper management well under way, advanced data-tracking and communications systems will be far more effective. Without the efficiencies introduced by wide participation and shared responsibility, a manufacturer might be overwhelmed by the volume of information provided by the new technologies. As with most aspects of the new manufacturing environment, organizational and technological changes complement each other and cannot be separated without tremendous costs in corporate effectiveness.

Effective decision making depends on trained personnel. Employees not only must be competent in their own jobs, but also must understand the relationship of their jobs to overall corporate operations and goals. Developing the required knowledge will require extensive training, by a variety of mechanisms and a significant amount of job rotation at every level of employment. As new technology is implemented, substantial training can be expected from equipment vendors, but the broader scope of job responsibilities is likely to require off-site classroom courses, company-wide seminars, and on-the-job training. Training will need to be extended to every employee as a corporate necessity, and it will account for a much higher percentage of working time and total costs than it has traditionally. Job rotation also can be expected to be far more extensive than has been common in U.S. firms. Short-term efficiency will be lost to a degree as employees rotate into unfamiliar jobs, but the long-term benefits, in terms of systems knowledge and a greater sense of a corporate team, will be indispensable for competitive operations.

EMPLOYMENT SECURITY

Creating a company-wide environment suited to competitive production with advanced technology will require a strong commitment to employment security—the ability of a firm to retain its employees even though changing market conditions or advancing technology may significantly change the content of the jobs they do. There is little doubt that manufacturing will no longer be a strong source of employment for the unskilled and semiskilled. Factories will employ fewer people in these groups in particular, but the number of people employed among skilled workers and managers also will decline. The remaining jobs, however, will be challenging, rewarding, and in demand. Competition for those jobs and competition for good people will make strong job security an interest of both employee and employer.

Information systems, training programs, and changes in organizational structures will represent huge investments in human capital. Cyclical layoffs or unnecessary turnover would severely limit the return on that investment and risk a complete breakdown in the company's operations. Employment security is crucial to engendering commitment of workers as true stakeholders. Increased responsibility and participation will improve the attractiveness of manufacturing jobs, but employees are not likely to feel a strong stake in the company unless they believe it has a stake in them. From the perspective of both company and employees (unionized or not), job security is a critical principle to pursue, which in itself will represent a significant change in the attitude of management and labor.

Although various mechanisms will be used in pursuit of a stable work force, absolute job security is likely to remain both elusive and a source of contention. Some companies recently have been very successful in relocating unskilled and semiskilled workers to other plants and in retraining them to perform new and varied tasks in the automated factory. Many employees, however, will be unable to adapt to a new environment that requires more skills, knowledge, and responsibility. Following this shakeout, however, it will be feasible and advantageous for employers to provide strong job security for a core group of employees. This core group would be capable of handling daily operations, and subcontracted temporary workers would be hired to meet surge demands. These temporary crews would perform specific duties that do not require extensive knowledge of the company's operations. They would be managed by the operatives usually responsible for those tasks to maintain continuity in decision making. This approach will insulate the core staff from fluctuations in market demand (essentially making the core labor a fixed cost), will provide employment opportunities for previously displaced workers, and, if widely used, may change the nature of unemployment trends in the macroeconomy. The approach is already used extensively in the airframe industry.

Job security also may be strengthened by the trend to perform previously subcontracted work in-house, although this trend will vary across industries and firms depending on size, available technology, and product mix. No company can do everything well, and, in many industries, subcontracting is a way to share risks, costs, and expertise. In some industries, however, advanced process technologies are likely to provide sufficient capacity and flexibility to encourage firms to pull subcontracted production in-house. The advantages in quality control, production scheduling control, and design change would reinforce the job security benefits of such a strategy. In fact, the trend can be seen already in the domestic mainframe computer industry. In other industries, the advantages of small, focused factories may create more subcontracting than has been traditional. Despite these variations, many companies will find that the advantages of in-house production outweigh the disadvantages, particularly in maximizing return on the large investment in human capital.

INCENTIVES, EVALUATIONS, AND DECISION CRITERIA

Traditional measures of success in manufacturing will be inadequate for tomorrow's manufacturing environment. New measurements, as well as new rewards, will be needed to manage production effectively, maintain employees' motivation, justify new investments, and stimulate stockholders' interest.

In the area of factory operations, managers will need new criteria on which to base operational decisions. Mechanisms for improving factory effectiveness, such as precise inventory control systems, material requirements planning, in-house production of previously subcontracted work, production process planning, and the many aspects of factory automation, will change the operations of the future factory. The criteria that managers have traditionally used to make operational decisions will change, and, in many cases, the decisions will change. A factory using just-in-time inventory control, for example, will have less input inventory on hand than a manager may have been accustomed to having. As another example, changes in production processes and work flows will change the criteria used to judge effective machine utilization rates, manning levels, acceptable work-in-process inventory, and tooling requirements. Managers will need retraining to alter their thinking about the effective operation of the factory to prevent old habits from inhibiting potential cost savings, quality improvements, and overall effectiveness.

Evaluations of individuals—both managers and workers—are likely to be much more subjective than they have been traditionally. Quantifiable improvements in individual performance, such as increasing output per hour or shift, will not be applicable to automated, integrated production with emphasis on project teams. Objective indicators of performance will remain in areas such as quality, delivery, process system costs, customer satisfaction, and company earnings, but these will reflect more on group efforts or the total work force than on individuals. Consequently, individuals' pay is expected to shift from hourly wages to salaries; pay will entail a greater emphasis on bonuses based on improvement in short-term results, long-term improvements in the total system, and achievement of the goals of that particular level of the organization. The "profit center" and "cost center" focus used in the past as a basis for judging individual performance can be expected to be replaced by a systems focus. Subjective assessments of individuals' skills and competence by their peers will affect salary decisions at least indirectly. Promotions will remain a form of recognition and opportunity for increased responsibility, but increased use of project teams and job rotation is expected to diminish the importance and obvious benefits of promotions; the elimination of most middle management positions will reinforce this trend.

At the company level, evaluation of performance will depend largely on a meaningful management accounting system.[2] Traditional methods that aggregate data, allocate costs based on direct labor, and compute data over long intervals (usually monthly) will be ineffective and counterproductive in the new environment. With computer-integrated manufacturing technologies, companies will be able to measure their performance continually by resource category, cost center, and product class and will abandon cost accounting systems based on direct labor. New accounting systems will give manufacturers the accurate, timely data they need to respond rapidly to changing conditions. The availability of more relevant data will give almost everyone in the business a clear perspective on total performance and its response to key decisions on matters such as investment, personnel, subcontracting, and research and development expenditures. (Appendix B gives a full analysis of the changes needed in cost accounting systems and the capabilities offered by new manufacturing information systems and computer integration.)

Changes in accounting procedures will contribute to the strong trend toward balancing short-term results against long-term prospects in determining the health of a manufacturing firm. New criteria will be developed to give stockholders and investors a basis for assessing the steps a firm is taking to ensure its long-term competitiveness. These criteria may include research and development expenditures, the amount and kinds of investment over a given period, training and recruitment patterns, and the activities of major competitors. None of these indicators will be conclusive, but as a package they will give stockholders more information than is common currently.

FUTURE FOCUS

In the new manufacturing environment, efficiency alone will not ensure success. Foresight will be the ultimate competitive weapon, because market share and profit margins are likely to be small for the followers. A long-term, future-oriented focus, extending beyond the next quarter or year, will be a competitive necessity. Manufacturers will need to devote an increasing amount of time, money, and energy to those parts of the business that will have a preponderant impact in the future, particularly product and process research and development.

The pace of change is expected to be rapid, so emphasis on strong in-house research and development will be a necessity for firms seeking a leadership position. Manufacturers will need to accept the risks inherent in long-range research. Investments in scientific and engineering personnel, laboratories, and computers are expected to be a significant portion of total capital budgets. At the same time, the need to implement new technologies, introduce new products, and attract talented personnel will be expensive. Some companies will share costs by participating in research consortia, which in fact may be the only viable method of research in some industries. Other companies will save research costs by using licensing agreements, but as product life cycles shorten, the value of licensing as a relatively inexpensive way to enter new markets can be expected to diminish. Companies will face difficult choices in striking a balance between spending for future and immediate competitiveness. Similar circumstances exist today. The major difference is that future manufacturers will probably have much less ability to milk profits from new products because competing entrants will be close behind. The costs of being a follower will be more apparent, so the weight given to future-oriented investments should be much greater than it has been traditionally.

CONCLUSION

With these changes in the human and organizational components of manufacturing, the factory will become a much different factor in society. Although opportunities for unskilled or semiskilled labor will diminish, the jobs that will be created are expected to be challenging and of high quality. Also, manufacturing jobs will be in demand among graduate engineers, who do not generally prize them today, and there may be too few to go around. Firms will have such large investments in people that they will make extraordinary efforts to retain employees, which will limit job creation at existing plants. This constraint may be countered somewhat by the trend in some industries toward microfactories, although the labor requirements of such facilities may be quite small. Employment opportunities also will arise in industries producing goods yet to be invented and in the variety of services that can be expected to develop to support future manufacturers.

These changes in the factory will permeate the social and economic fabric of the nation. Changes in internal factory operations will affect relations with unions, subcontractors, wholesalers and retailers, producers of services, and other economic activities outside but closely related to factory operations. The expectations and opportunities of workers at all levels will be affected by the cultural revolution that has already begun in manufacturing. For many companies, the rate and direction of change will be determined through the collective bargaining process; for other firms, less formal approaches of labor-management cooperation will be used. None of the changes will be sudden, however, and no two industries will progress at the same pace.

In fact, there may very well be a backlash from both managers and workers who have a strong stake in traditional relationships and organizational structures. Consequently, the changes in people and organizations will, at best, proceed in fits and starts, but the benefits in terms of manufacturing effectiveness and profitability are expected to be so clear that these difficult cultural changes will be implemented. The specifics of these changes are difficult to predict because they are based on individual decisions in a vast variety of circumstances. The direction of change, however, is becoming increasingly clear, and the repercussions will be wide-ranging.

NOTES

[1]The findings from field work conducted for a Manufacturing Studies Board report on effective practices in implementing advanced manufacturing technologies have been consistent with these predictions. See Committee on the Effective Implementation of Advanced Manufacturing Technology, 1986, Human Resource Practices for Implementing Advanced Manufacturing Technology, Washington, D.C.: National Academy Press.

[2]Kaplan, Robert S. 1985. Accounting Lag: The Obsolescence of Cost Accounting Systems. Pp. 195-226 in Clark, Kim B., Robert H. Hayes, and Christopher Lorenz, eds. The Uneasy Alliance: Managing the Productivity Dilemma. Boston, Mass.: Harvard Business School Press.

4
Considerations for Government

While the technological and organizational changes needed in manufacturing will be the result of market pressures and private initiatives, government policies will play an important role in setting the economic and political environment for private business. Workers, managers, and technologists will develop and implement the strategies necessary to have a competitive manufacturing sector in the future. Their acceptance of change, pursuit of new breakthroughs, and willingness to take risks will determine the long-term competitiveness of U.S. manufacturing. Government policies can stimulate this process, and every tax, spending, and money supply decision has an impact, but it is very difficult to identify specific policies that would be indisputably beneficial and politically acceptable across the broad spectrum of U.S. industry. Reports from both public and private groups have made specific policy recommendations (see Appendix C). The value of adding a set of similar recommendations, or even of endorsing what others have said, was thought to be negligible.

The magnitude of the changes that the Manufacturing Studies Board (MSB) is forecasting for U.S. industry, however, should have an impact on the way policymakers think about manufacturing. Policymakers will need to recognize that the policies used in the past to help domestic manufacturers may no longer provide the desired results and that the policies used by government agencies as consumers of manufactured products may no longer be effective. For instance, if the changes needed are as profound as this study indicates they will be, many companies in many industries will be unable or unwilling to adapt. These firms will want to maintain the status quo and undoubtedly will petition government to slow or stop the process of change. Calls for trade protection, tax relief, or direct subsidies may result in policies that work against the long-term goal of a competitive manufacturing sector and are detrimental to the long-term health of the firms or industries being helped. Experience demonstrates clearly that in all but a very few cases, such government intervention has slowed change as intended, but has also damaged the interests of consumers and other industries and has not helped the long-term prospects of the industry assisted.

Government assistance may sometimes be unavoidable, but it should be contingent on explicit commitments by the industries involved to make the changes necessary to regain their competitiveness. Such a quid pro quo for government assistance would create the correct impression within the affected industry that change is mandatory and that the industry itself must devise and implement the necessary strategies. Government policies can help ease the dislocations suffered by various industries and regions by using a process that stimulates future growth rather than preserving the status quo.

Demanding an explicit quid pro quo in return for trade protection, loan guarantees, regulatory relief, or other government-provided support represents a departure from past government practice. Other government initiatives to support manufacturing, however, need not require dramatic changes in policy. Existing programs provide unemployment compensation, training and relocation assistance, trade adjustment assistance, and protection against dumping by foreign firms. These programs may receive inadequate funding or include too many restrictions to provide sufficient support in a changing environment, but they continue to represent important government support mechanisms. Federal activities in transportation, education, research, and defense directly benefit the competitiveness of U.S. manufacturers. These and many other federal efforts contribute to the economic and social infrastructure in a way that ensures that all firms are affected fairly and that progress and success depend on market factors, not government.

These types of supports should continue to be the major priority in government policy toward manufacturing. In addition, several areas of traditional federal concern that will be affected by the coming changes in manufacturing may need to be reevaluated to ensure the continued efficacy of government programs. The areas are trade, education, research, and defense.

TRADE

Continued progress in international economic development and the growing internationalization of the U.S. economy will create increasingly strong competition in manufactured goods from an expanding number of foreign producers, even in industries that traditionally have not had significant import competition. These developments will change the international and domestic trading environment. Two specific developments should be of particular interest to policymakers.

First, while an open and fair international trading system should continue to be the goal of U.S. policy, it may be increasingly difficult to devise effective trade policy, in part because of the diminishing number of clear-cut trade policy tools. Flexibility in U.S. trade policy will be hampered by such factors as the growing number of foreign manufacturing facilities in this country, the increasingly complex pattern of equity ownership across national boundaries, a growing incidence of foreign products in the product lines of domestic manufacturers, and a trend among multinational companies to make components in scattered plants and assemble them at a single location. It is becoming increasingly difficult to determine what is an American firm and just what is meant by "U.S. manufacturing." In such an environment, any trade measure designed to benefit domestic producers or encourage other countries to open their markets not only will result in additional costs to domestic consumers but also may have conflicting effects on a single industry or even a single company. Developments among private firms, specifically the integration of international production facilities, may increasingly preempt the ability of the federal government to make a substantial positive difference in trade.

One exception to this general trend, however, should be noted. Government protection of intellectual property rights will be an important and essential support for U.S. companies competing in both domestic and foreign markets. Too many foreign companies have used infringement of patents, trademarks, and copyrights to establish market share, sometimes with shoddy goods that damage the reputation of the original manufacturer. Effective enforcement across national borders has proven difficult, particularly when foreign governments are slow to recognize the problem, and delays compound the damage. Court proceedings can sometimes award restitution and legal sanctions against continued infringement, but such "solutions" are too time-consuming and often temporary. U.S. manufacturing companies and U.S. consumers have a tremendous stake in this issue because together they are the major victims. Provisions in the 1984 Trade Act make benefits under the U.S. general system of preferences conditional on the protection of copyright by the importer of American works; this initiative already has begun to have an impact. The U.S. government must remain committed to continued vigilance and the implementation of strong sanctions to ensure quick and effective resolution of international patent, trademark, and copyright infringements.

A second development in the trade area is that U.S. firms that have had no experience with foreign trade will be thrust into the international economy as imports compete in a growing range of products and markets.[1] In response, many manufacturers will adopt new technologies and management techniques and in most cases will compete effectively against imports in the domestic market. Many firms, however, may not consider competing in export markets unless they receive special encouragement, as well as help in securing expertise and information. Government programs already exist to provide this assistance, and private initiatives can be expected to meet many of these needs, for example, through export trading companies, but policymakers should be aware of the need to expand the export base as much as possible and recognize the potential for additional government efforts in ensuring access to foreign markets and encouraging export activity.

EDUCATION

The reduction of direct employment opportunities in manufacturing and the different skill requirements of future manufacturing jobs will put new demands on postsecondary training and education. Technical schools will need to base their training for displaced workers and young people entering the labor force on realistic reassessments of industry's skill requirements, recognizing that industry's ability to predict its needs sufficiently in advance to accommodate educational planning cycles is weak at best and that most future manufacturing jobs will not be on the factory floor. Furthermore, skill requirements for the remaining positions in manufacturing may vary significantly among plants, creating a need for customized training programs. Cooperation between public educational institutions, such as technical and vocational schools, and private training programs can be expected to increase, but there also may be increasing policy debate over the distinction between the responsibilities of public education and private training needs.

At the university level, manufacturing will compete with other sectors for broadly educated engineering graduates, and rapid technological change will make it more necessary for older engineers to update their skills. The conflicting pressures of a growing demand for knowledgeable engineers and a rapidly expanding knowledge base will increasingly strain the ability of universities to supply enough engineers with broad-based knowledge. A balance is needed between good grounding in one field and interdisciplinary instruction. That balance is likely to improve as industry learns how to articulate its needs and as salaries reflect those needs.

A shortage of engineers would pose a serious barrier to progress in manufacturing and many other fields. There is continuing debate about the likelihood of a shortage of engineers—some observers claim a shortage already exists. Part of the debate stems from the relative decline in the number of U.S. students in postgraduate engineering programs. This drop seems to reflect industry's high demand for baccalaureate-level engineers and the small perceived benefits of a postgraduate degree. If the situation is detrimental, industry should work with universities and government to provide incentives for postgraduate study and adjust compensation levels to encourage more postgraduate work. Of course, an overall increase in engineering students would increase the pool of potential graduate students, so efforts should be made to increase the number of students entering engineering. This is a complex problem, requiring adjustments on at least two levels. First, students leaving high school will need a thorough grounding in mathematics and science so they will be interested in and not intimidated by engineering curricula. Second, relative starting salaries must be adjusted to encourage bright students to enter engineering instead of other lucrative fields. Both of these adjustments require a commitment by society in general that manufacturing is important and a desirable career choice.

In management education, future business programs at both the undergraduate and graduate levels will need to reflect the extensive organizational and technological changes on the factory floor. For example, the criteria for operational decisions are likely to change significantly, which management education must reflect. The relationships between individuals and functions in the factory environment also will change. Graduates will need realistic expectations to become effective in the new environment. Provisions also will be needed to educate current managers about new production processes, strategies, and goals to change their traditional approaches to factory management. As with worker retraining, company-provided training and public education will need to work in partnership to produce effective manufacturing managers.

RESEARCH

A major advantage of U.S. manufacturing has been in basic research and the resulting product and process applications. Although the nation continues to spend far more than other countries on research and development, the U.S. share of world research spending has declined steadily in the past 20 years as other countries have increased their expenditures. The United States can no longer be assured of unchallenged leadership in research findings and applications. This shift could prove crucial to manufacturing competitiveness.

Manufacturing is becoming increasingly science based; that is, scientific knowledge is virtually a prerequisite to the effective design and implementation of advanced process technologies. The integrative nature of future manufacturing technologies (see Chapter 2) is less tolerant of imprecise data than traditional manufacturing operations. The "art" of a skilled machinist, for example, cannot be duplicated exactly by an automated machining center, especially for a complex part. The machine must "know" what is happening at the interface of the part and the cutting tool, which varies tremendously with the material, size, and shape of both the part and the tool. Enough is known about the various interactions to automate the basic machining process, but the fundamental science is poorly understood. It will need to be known in detail as tolerances get tighter and the number of available materials and processing technologies changes. Just to program this one machining application involves thermal dynamics, materials science, surface physics, and a number of other disciplines; given the huge number of process activities in the factory, the need for scientific research explodes.

The required level of investigation will be far more precise than it has been traditionally. Even now, research into the surface behavior of materials is being conducted at the atomic and subatomic levels. These and other research efforts require sophisticated equipment, controlled environments, well-trained personnel, and time—costly but necessary inputs for progress. The ability of supercomputers to simulate physical phenomena may help control costs, but actual experimentation and measurement cannot be replaced completely.[2]

As these developments unfold, federal support of basic research in both government laboratories and universities will be increasingly important to the health of U.S. manufacturing. Foreign competition, more sophisticated research, and the burgeoning need for scientific knowledge in many aspects of manufacturing will place growing demands on federal research funds. Policymakers will need to recognize the increasing importance of research in the long-term competitiveness of manufacturing and allocate resources accordingly.

DEFENSE

Rapid technological progress in manufacturing in defense-related industries will be even more of a cornerstone of defense policies than it has been in the past. For a number of historical reasons, the U.S. defense posture has been based on technological superiority of weapons systems, which depends increasingly on sophisticated manufacturing processes. Advances in manufacturing technologies will provide broad benefits in terms of the ability to design and manufacture increasingly complex weapons systems. The technologies will bring higher quality at lower cost with more cost-effective customization capabilities and better price-performance ratios (see Chapter 2 and Appendix A). Advanced technologies will give defense contractors the responsiveness, flexibility, and cost effectiveness necessary to meet a broad range of weapons design requirements and production schedules. The new manufacturing processes made possible by new technologies also will result in completely new products. Defense officials therefore have a clear interest in stimulating the implementation of advanced manufacturing technologies and organizational structures in the defense industrial base. Because defense contractors often respond to different market signals than their commercial counterparts (even in the same company), federal officials have a difficult problem ensuring that defense procurement policies give contractors strong incentives to implement new process technologies.

Because of the inherent problems in defense procurement procedures, defense officials have used specific programs to encourage manufacturing development and implementation by contractors. The Manufacturing Technology programs, Industrial Modernization Incentive Program, and the Technology Modernization programs have used different strategies and criteria to support manufacturing technology improvements by defense contractors. The effectiveness and necessity of these programs have been debated and their funding levels have fluctuated, but they remain the only specific government programs directed at improving manufacturing technology. These programs reflect the importance of advanced manufacturing technology in weapons systems production and the shortcomings of the defense procurement system in providing incentives for manufacturing process improvements. Policymakers will need to recognize the growing link between advanced manufacturing technology and advanced weapons systems and address ways to provide incentives for manufacturing process modernization, either through major corrections in the procurement process, consistent adequate funding for focused programs, or a combination of both.[3]

CONCLUSION

Government should continue to help companies and industries unduly hurt by the rapid change in manufacturing. It should help primarily by continuing to provide infrastructural support to manufacturers and their workers and secondarily by easing the negative impacts of the many changes expected in manufacturing. Policy should support and encourage the emergence of a technologically advanced, competitive manufacturing sector through continued strong infrastructural programs. In this vein, this report has three broad suggestions for future government policies toward manufacturing.

- Government initiatives to help special interests adversely affected by change should (1) secure explicit commitments from the industries affected to take the steps necessary to regain competitiveness and (2) ease the short-term economic and social dislocations without disrupting a fair competitive environment for other producers and without hindering continued progress for U.S. manufacturing as a whole.
- Government programs should help speed adjustments and provide the necessary infrastructural support to manufacturers without undue government interference. Private-sector initiatives will be most effective in developing and implementing the changes needed to make U.S. manufacturers competitive.

- Government policymakers need to understand that the process of change in manufacturing will result in corresponding change in a number of areas in which government has primary responsibility, such as trade, education, research, and defense. Policymakers need to recognize these changes in the environment and adjust their policies accordingly to provide maximum service to the private sector and to achieve maximum benefits for government programs.

NOTES

[1]The National Association of Manufacturers estimates that 70 percent of American manufacturing is already confronted with import competition.

[2]For a full discussion of the use of computers in design and manufacturing see Committee on Science, Engineering, and Public Policy, 1985, Report of the Research Briefing Panel on Computers in Design and Manufacturing, pp. 214-235 in New Pathways in Science and Technology: Collected Research Briefings 1982-84, New York: Vintage Books.

[3]An MSB committee is currently studying the role and effectiveness of the Manufacturing Technology programs. Their initial findings are contained in a Phase 1 report. See Committee on the Role of the Manufacturing Technology Program in the Defense Industrial Base, 1986, The Role of the Department of Defense in Supporting Manufacturing Technology Development, Washington, D.C.: National Academy Press.

5
Summary

Manufacturing has already entered the early stages of revolutionary change caused by the convergence of three powerful forces:

- The rapid spread of manufacturing capabilities worldwide has created intense competition on a global scale.
- The emergence of advanced manufacturing technologies is dramatically changing both the products and processes of modern manufacturing.
- There is growing evidence that changes in traditional management and labor practices, organizational structures, and decision-making criteria are needed to improve the effectiveness of manufacturing operations, provide new sources of competitiveness, and introduce new strategic opportunities.

The effects of these forces are already being felt by the U.S. manufacturing community. Domestic markets that were once secure have been challenged by a growing number of foreign competitors producing high-quality goods at low prices. New technologies are helping U.S. manufacturers compete, but many technical and social barriers remain before advanced technologies have a major, widespread impact on manufacturing operations. Unfortunately, foreign competitors may well have overcome some of these barriers first, using new technologies to increase their competitiveness.

As these points indicate, the three trends now affecting manufacturing are closely interrelated. Increased competition has demonstrated the need for U.S. manufacturers to reexamine traditional human resource practices and their use of new product and process technologies. Corrective measures, however, cannot focus exclusively on either area, since technology will not be effective without changes in human resource practices, and the benefits from those changes are limited without the productive thrust offered by new technologies. Meanwhile, the competition intensifies, current production must be maintained, and the resources available to make the required changes always seem inadequate.

All of this poses a difficult dilemma for manufacturers who have depended on stability to maintain competitive production. Many manufacturers recognize the need to adapt, but do not know what changes are necessary or how to implement them. More than anything else, the key problem is that the forces affecting manufacturing require that managers think and act differently to bring about change in a systems context and that workers accept new roles and new responsibilities.

The major roadblocks to more competitive U.S. manufacturing are in the attitudes, practices, decision-making criteria, and relationships of both managers and workers. Chapter 3 described the kinds of practices that are likely to be required for future manufacturing competitiveness. That vision means that hierarchical, adversarial management structures will handicap attempts to improve competitiveness. Employees at all levels of the organization will need to be viewed as a resource, and the organization will need to be structured so that everyone will have the opportunity and responsibility to make the maximum contribution. Furthermore, the importance of the manufacturing function in the total corporate context will need to be recognized. Functional integration based on a clear understanding of the manufacturing systems concept will be a major key to competitive success. This way of thinking about manufacturing is foreign to most managers, workers, and educators in this country, and it may be overly optimistic to expect such a dramatic shift in attitudes and culture. Ingrained attitudes will be difficult to change and may require a generational shift.

This report has tried to provide some direction, not a solution. Circumstances vary too much to try to prescribe specific actions, but the direction for change should be clear. The use of new advanced manufacturing technologies is insufficient. The key is to focus on evaluating traditional managerial practices, relationships, decision-making criteria, and organizational structures to determine specific strengths in responding to competitive pressures. The renewed organization will be in a better position to implement new technologies and further strengthen competitiveness. For some companies, however, attempts to implement new technologies will force labor and management changes. Managers will need to realize that implementation of advanced manufacturing technologies to automate existing processes will yield suboptimal results. Efforts to optimize the technologies will demand creative thinking to take advantage of the opportunity to redesign many processes, simplify many designs, and change the flow of work on the factory floor. This creative thinking and the necessary cultural changes will be the major obstacles to attaining improved competitiveness.

Government can play a limited role in encouraging and supporting the changes in manufacturing, but the impetus must come from private companies. In general, the main responsibility for government is threefold: (1) to recognize the importance of a strong manufacturing sector as a source of goods for international trade and as a crucial factor in continued economic prosperity and strong defense; (2) to support the process of change in manufacturing; and (3) to stay abreast of the changes taking place in manufacturing and adapt government policies and programs to maximize their effectiveness in the new environment. In addition, some specific government activities, for instance in education and research, will need to be particularly sensitive to manufacturing requirements and ensure that necessary resources remain available.

To summarize, U.S. manufacturers are facing a crucial challenge. Traditional markets are being attacked by imports and traditional practices are not producing adequate results. Changes in labor and management attitudes, organizational design, and the role of the manufacturing function in the total corporate system are needed to regain and maintain competitiveness. New technologies will help this process, but manufacturing strategies will need to be evaluated to ensure both that the right technologies are used and that the full potential of those technologies is realized.

U.S. manufacturing is on the threshold of an exciting new era—the challenges are daunting but the opportunities are unprecedented.

APPENDIX A
The Technology of Future Manufacturing

Although all the interrelationships and long-term implications of advanced manufacturing technologies are not yet well understood, the direction of future developments is relatively clear. This appendix describes the technologies that are likely to have a major impact on manufacturing competitiveness, indicates the ways in which those technologies interact, identifies additional research needs, and discusses some of the issues that are likely to be encountered in implementation. The technologies have been divided into materials, material handling, material transformation, and data communication and systems integration. These categories are highly interdependent, and divisions are not always distinct, but this categorization provides an effective structure for a broad overview of the major technologies.

DEVELOPMENTS IN MANUFACTURING MATERIALS

Materials developments have a substantial impact on manufacturing in both product design and process engineering. New products can require different materials and materials processing, and new materials themselves often spur new products and new process development. Many of the materials developments in manufacturing do not involve new materials, but rather substitutions, upgradings, and new concepts for conventional materials. Pressure to lower costs and raise product quality has led to some major shifts in materials selection. The consideration of doubly precoated steel for corrosion resistance in automobiles is an example. Developments in both conventional materials and new materials are equally significant to future manufacturing.

This section will focus on materials developments that are ready for manufacturing implementation, with minimal emphasis on research systems that have had little technology transfer effort. Developments in metals, polymers, ceramics, and glasses will be considered, followed by a brief discussion of some emerging issues that should be brought to the attention of policymakers.

Metals and Metal-Based Composites

Major developments will continue in the processing of conventional metallurgical systems. For example, large tonnages of carbon and stainless steel sheet and strip will be continuously cast. Similar developments are certain in nonferrous alloy areas as well. While the continuously cast products will have some minor metallurgical variations to be considered, the major impact will be economic, allowing the basic metals industry to remain competitive in many sheet and strip applications.

Increased use of warm- and cold-formed steel parts can be expected, with emphasis on near-net-shape processes to save metal and avoid intermediate processing steps. Similar forces will continue to drive powder metallurgical processing, although it must be emphasized that advances in precision forming and powder metallurgy have been slow over many years rather than a sudden breakthrough. Powder processing will be facilitated as cleaner powders reach the market. Powder metallurgy produced tool steels continue to offer advantages over conventional tooling stock.

Superplastic forming will continue to increase in aerospace manufacturing; some initial applications have occurred in the B-1 bomber program. Increased market penetration will entail major changes in tooling and manufacturing technology. Thermomechanically processed 7000 series aluminum alloys are available for such forming, as are aluminum-lithium alloys. However, little of these alloys are available from domestic sources; most of this material is being obtained from the United Kingdom.

Many metal-processing alternatives will be examined to facilitate in-line processing systems. Improvements can be expected in the control of metal structure, with increasing awareness of the importance of grain orientation (texture), residual stress, and surface quality. In steel surface treatment, increased use of induction or laser hardening can be expected as lower cost alternatives to carburizing. Laser welding is seen as growing in the auto industry, perhaps at the expense of electron beam techniques. Power systems manufacturing may turn to welded rotor fabrication to allow increased use of attractive alloy combinations while sidestepping large forging development problems.

Many cases of metals substitution can be expected. The upgrading of alloys in small parts should increase product quality, reliability, and processing response without grossly increasing overall metal procurement cost. Aluminum can be expected to continue to replace copper in many heat exchange applications. Silicon-based switches are expected to replace iron-based magnetic devices. Also in the electronics industry, changes in plating metals are expected, with gold giving way to palladium-nickel and iron. Some observers see increased use of molybdenum-based alloys, particularly as new developments solve some of the traditional corrosion problems. Production of these alloys is energy intensive, however, and the domestic supply is limited.

In the new metals area, continued progress is expected in the development of metal matrix composites, particularly using metals such as aluminum and magnesium reinforced with silicon carbide. Three major types of reinforcement are receiving particular attention: continuous monofilament, discontinuous, and continuous multifilament yarn. Each reinforcement requires a specific fabrication process, including diffusion bonding, hot molding, power blending, forging, casting, pultrusion, extrusion, and rolling, often in combination. Many of the material defects and anomalies that have plagued metal matrix development are attributable to the manufacturing process. As experience in these processes builds and the manufacturing technology evolves, application problems caused by material defects should decrease. In aerospace structures, for example, the application of specific metal matrix materials varies according to expected environment, design loads, stress, and temperature variations. Designers of flaw-critical parts select materials for their strength and ductility as well as their resistance to crack growth. Unfortunately, research on fracture mechanics in the area of metal matrix composites is much less developed than for polymer matrixes and has not been widely disseminated for use by designers.[1]

Progress in reducing material defects and continued application experience will result in broader applications for metal matrix composites. For example, automobile manufacturers are gaining experience with aluminum-silicon carbide in piston ring and crankshaft applications. Power systems applications are seen for nickel superalloy and stainless steel matrix composites strengthened with silicon carbide. Such systems allow increases in elastic modulus and desirable decreases in the coefficient of thermal expansion.

Much research has focused on rapidly solidified metals and amorphous metals. Introduction of such materials into the manufacturing sector is expected to be slow, with the major exception of the iron-boron-silicon-carbon system being broadly used for distribution transformers. Slow emergence is also predicted for nickel and titanium aluminides despite extensive research. Recent aluminide development has greatly increased its durability, and some jet engine applications can be expected.

Polymers and Polymer-Based Composites

Polymers and polymer-based composites will probably continue to displace carbon steel and aluminum in a significant segment of structural and paneling applications. This trend may be most prominent in the automobile industry, but it is also likely in electronic hardware, appliance chassis applications, and home building components. Current automobiles contain about 157 pounds of plastics and polymer-based composites. By 1995, this should grow to 213 pounds as polymer materials are used increasingly in body panels. The increased use of plastic is expected to save motorists about $200 per year through fuel economy, corrosion resistance, cheaper repairs, and lower insurance rates.[2]

The applications seen in models such as the Pontiac Fiero will spread for reasons of economy and safety. In the Fiero, the horizontal body panels are made of flexible glass fiber-filled polyester (a sheet molding compound) and the vertical panels are made of relatively stiffer glass-reinforced polyurethane (a reaction injection molded product). Beyond the body panel substitutions, further use of polymers can be expected in the automobile structure. For example, polymeric leaf springs have been used in some automobiles since 1981. While glass-polyester and glass-polyurethane materials will see major tonnage applications, future use of polymers in automobiles can be expected to rely on reaction injection molded thermoplastics as well. Beyond these applications, the expanded use of coatings (including paint) on steel can be regarded as an area where polymers will intrude further into the sheet metal markets.

The growth of polymer panels and structural shapes has necessitated the development of adhesives as a joining medium. In addition to the increased use of conventional adhesive materials, advanced work is under way on adhesive systems for higher-temperature applications (epoxies, polyimides), adhesives with greater strength and elastic range than epoxies, primerless adhesives (silicones perhaps), and faster-curing adhesives (cyanoacrylates, urethanes, etc.). In some instances, adhesive development overlaps sealant systems (silicones). Beyond the relatively simple polymer applications, adhesives are increasingly required for bonding dissimilar materials, particularly when differential thermal expansion must be accommodated. For example, rivets cannot be used to join plastic liners to metal trailer bodies because of differential thermal expansion. In some metal joining developments, adhesives are being used in conjunction with spot welding to replace riveting. However, it must be emphasized that the use of adhesives for nonpolymeric joining has been slowed considerably by concerns about reliability, contamination, degradation, and consumer acceptance.

Another major shift in polymer materials use can be foreseen where flame retardation is a dominant consideration. Underwriters Laboratory interpretations of smoke, flame, and toxicity requirements are leading to shifts away from polyvinyl chloride-based systems to fluorocarbons.

High-Technology Ceramics

Advanced high-technology ceramics[3] are nonmetallic materials having combinations of fine-scale microstructures, purity, complex crystal structures, and precisely controlled additives. In contrast to traditional ceramics, which are made from natural raw materials such as silica and clay, advanced ceramics are made from artificial raw materials, such as aluminum oxide, zirconia, yttria, silicon nitride, and silicon carbide, which are formed, sintered, and treated under precisely controlled conditions. The advantage of such fine ceramics is their ability to play both functional and structural roles. Functional uses include optical devices, motors, transducers, sensors, and semiconductors; structural uses include those that require high specific strength, high wear resistance, and high corrosion resistance. Both roles will be increasingly important in manufacturing applications.

Currently, high-technology ceramics are used most often in electronics, including optical fibers, multilayer ceramic-to-metal interconnecting and mounting packages for integrated circuits, ceramic multilayer chip capacitors, piezoelectric ceramic transducers, and chemical, mechanical, and thermal sensors. Processing for these applications is generally an extension of standard ceramic technology, in which powders are pressed or formed with binders and sintered to densify the ceramics. Incremental progress has improved results, but major improvements cannot be expected until semiconductor processing techniques are applied to ceramic components. Techniques such as selective-area ion implantation and laser-induced recrystallization will greatly improve many of the properties of electronics ceramics.

High-technology structural ceramics are used as coatings and for monolithic and composite components. Major applications include tooling for metal working, wear components in a variety of abrasive environments, bioceramics for bone replacement, and military ceramics for radomes and armor. Major efforts are under way in both the United States and Japan to use structural ceramics in a variety of automotive applications, including engine wear components, turbochargers, bearings, and a variety of diesel engine components.

Significant strides have been made in the mechanical properties and reliability of monolithic structural ceramics. Understanding of strength-limiting flaws and temperature-brittle fracture behavior has improved greatly, but further work is needed to improve reliability. Improvements are needed in powder synthesis, powder properties, near-net-shape fabrication methods, microstructure control, mechanical properties, and nondestructive testing methods. Important research also is needed to identify new, more complex ceramics.

Significant advances also are being made in thermal barrier coatings and ceramic matrix composites. Ceramic matrices combined with particulates, whiskers, or fibers of a different ceramic compound or metal have yielded composites with five times the resistance to fracture of the monolithic ceramics. Recent success has been reported in the use of metal ion implantation to reduce the relative friction resistance of ceramic diesel engine parts.[4] New research is needed to quantify the improved mechanical properties of composites, particularly fracture resistance.

In addition to the research required on the composition and properties of ceramic materials, much work is needed on the processing and product design requirements of ceramics. Promising directions in ceramic processing include the use of ultrafine powders and the use of chemical routes to supplement or bypass some of the powder-processing stages. Other requirements include the processing of fine-scale layered structures, processing of ceramic composites, joining of ceramic parts, and near-net-shape processing of complex parts to minimize machining requirements.

The rate of technical progress in ceramic materials and processing will determine the pace of commercial application. Major market penetration for structural ceramics depends largely on the progress made in automobile applications. Several Japanese firms have already introduced ceramic turbocharger rotors, piston rings, swirl chambers, and camshafts.[5] An almost totally ceramic engine is a major research objective of virtually every automobile manufacturer and should be extant by the early 1990s. As these automotive applications increase and the price falls, high-technology ceramics can be expected to see rapid growth in product and process (cutting tools) applications.[6]

General Issues Related to Materials

It is important to recognize the critical lack of data on and basic understanding of the physical properties of many materials. This lack is a severe handicap in manufacturing process development. Most materials handbook data have been generated for use in product design and service performance analysis rather than for process analysis. The lack of knowledge has become acute as software systems have emerged with powerful process control and process design capabilities. The requisite data inputs often involve combinations of stress, strain, strain rate, temperature, heat transfer, friction, and so on, which have not been studied, even for classic engineering materials.

This lack of data on manufacturing materials grossly compromises the effectiveness of computer-based process models, and it tends to foster undue reliance on the few materials for which an adequate data base seems to exist. Power systems in particular are plagued by a lack of materials innovation due to the awesome data base requirements for service performance analysis, manufacturing modeling, and code adherence. With the current low return on investment in much of the primary materials industry, little supplier information is being generated. While some manufacturers develop their own data bases, others find that largely empirical trials are the least expensive approach (from a local point of view).

The scientific community has been reluctant to get involved in data-generating efforts that involve "no new science." Federal funding agencies, reinforced by peer review systems, also have shunned this area. Progress in generating this data could have a significant impact on a variety of industries and process applications.

Another recurring concern is the frequent lack of domestic suppliers for new materials, such as some ceramics. Manufacturers are reluctant to begin using new materials systems when only one or perhaps no domestic supplier exists. Many new materials are initially imported in quantity from Japan or Europe.

A related problem is the growing tendency for manufacturers and wholesalers to limit inventory. Indeed, limiting inventories has emerged as a smart manufacturing practice, especially with high interest rates. However, this practice grossly limits the availability of new and many old materials for manufacturing trials. In fact, most of the materials in reference handbooks are not available in tryout quantities.

Lastly, there is an important interaction between the use of new materials and recycling and scrap practices. This is particularly the case in the automobile industry, in which a by-product of primarily steel construction has been the relative purity of car bodies as a source of steel scrap. The ease of recycling car bodies is being compromised by the materials substitutions now occurring, which could significantly increase material costs in a number of industries.

Although these problems slow progress, none is sufficient to prevent increased use of new materials in manufacturing if those materials are cost effective in production, performance, and maintenance. Limited availability is probably the greatest handicap, because the machining, tooling, and processing required for many new materials can be vastly different from those for traditional metal cutting; significant research in a production environment is a prerequisite for increased use. Fortunately, enough production experience is being accumulated with many new materials, particularly polymers, to demonstrate their advantages and to encourage efforts in other areas of materials research. Despite the handicaps, significant breakthroughs can be expected so that changes in manufacturing materials will keep pace with the many other developments on the factory floor.

MATERIAL HANDLING TECHNOLOGY TRENDS

This section will assess the major trends in material handling technology. Material handling systems are used to enhance human capabilities in terms of speed of movement, weight lifted, reach distance, speed of thought, sensory abilities of touch, sight, smell, and hearing, and the ability to deal with harsh environments. In this area, it is important to distinguish between equipment technology and design technology. Equipment technology is categorized by its primary functions: transporting, storing, and controlling materials.

Transporting

The material handling function of transporting material has been affected significantly by two trends—toward smaller loads and toward asynchronous movement. The former is the result of the drive to lower inventory levels through just-in-time production. It has been manifested in the development of numerous equipment alternatives that have been downsized for transporting tote boxes and individual items rather than the traditional pallet loads. The inverted power-and-free conveyor, powered by linear induction motors for precise positioning and automatic loading and unloading, is one example of the trend to develop transport equipment for small loads. Automatic guided vehicles (AGVs) for transporting individual tote boxes are also being developed, as are specially designed conveyors and monorails for tote box movement.

The use of asynchronous movement in support of assembly has existed for many years. For example, asynchronous material handling systems were prevalent in automotive assembly before the paced assembly line was adopted at Ford in the early 1900s, and the concept has been applied recently in some European automotive assembly operations. In the early 1970s, Volvo began using AGVs to achieve asynchronous handling in support of job enlargement. Asynchronous material handling equipment is often used to allow a worker to control the pace of the process.

The trend toward asynchronous movement appears to be partially motivated by the apparent success of Japanese electronics firms in using specially designed chain conveyors that place the control of the assembly process in the hands of the assembly operators. A workpiece is mounted on a platform or small pallet which is powered by two constantly moving chains. The platform is freed from the power chain when it reaches an operator's station.

After work is completed on the workpiece, the operator connects the pallet to the chain, and it moves to the next station; if the next station has not completed its work on the previous piece, the pallets accumulate on the chain.

Asynchronous alternatives include using AGVs as assembly platforms and for general transport functions; "smart" monorails for transporting parts between work stations; transporter conveyors to control and dispatch work to individual work stations; robots to perform machine loading, case packing, palletizing, assembly, and other material handling tasks; microload automated storage and retrieval machines for material transport, storage, and control functions; cart-on-track equipment to transport material between work stations; and manual carts for low-volume material handling activities.

Storing

The major trends in material storage technology are strongly influenced by the reduction in the amount of material to be stored and the use of distributed storage. Rather than installing eight to ten aisles of automated storage and retrieval equipment, firms are now considering one- and two-aisle systems. Rather than being designed to store pallet loads of material, systems are designed to store tote boxes and individual parts. Also, rather than a centralized storage system, a decentralized approach is used to store materials at the point of use.

Among the storage technology alternatives that have emerged are storage carousel conveyors; both horizontal and vertical rotation designs are available. Furthermore, one particular carousel allows each individual storage level to rotate independently, clockwise or counterclockwise. A further enhancement of the carousel conveyor is automatic loading and unloading through the use of robots and special fixtures.

A number of microload automated storage and retrieval systems have been introduced in recent years. The equipment is used to store, move, and control individual tote boxes of material. Rather than performing pick-up and deposit operations at the end of the aisle, the microload machine typically performs such operations along the aisle, since it is used to supply material to work stations along each side of the storage aisle.

Of particular interest has been the introduction of storage equipment for production applications that previously was used for document storage in office environments. The trend toward lighter loads has resulted in a shift of technology from the "white-collar" environment to the "blue-collar" environment.

Despite the apparent need for automatic storage and retrieval of individual items, few equipment alternatives are currently available, and those that are have not gained wide acceptance. This particular void in the technology spectrum has existed for a number of years and does not appear to be of current interest to material handling equipment suppliers.

Controlling

The ability to provide real-time control of material has elevated material handling from a mundane "lift that barge, tote that bale" activity to a high-tech activity in many organizations. The control aspect includes both logic control and the physical control of material. In the area of logic control, the ability to track material and perform data input-output tasks rapidly and accurately has had a major impact on material handling. In physical control, automatic controls have been added to a number of material handling equipment alternatives, allowing automatic transfer and assembly.

Perhaps the fastest-growing control technology today in material handling is automatic identification. Likewise, the expectation is that the greatest impact in the future will come from the application of artificial intelligence to transporting, storing, and controlling material. Among the alternative sensor technologies available to support automatic identification are a wide range of bar code technologies, optical character recognition, magnetic code readers, radio frequency and surface acoustical wave transponders, machine vision, fiber optics, voice recognition, tactile sensors, and chemical sensors.

The growth in the use of bar code technologies is due to three developments: bar code standardization, on-line printing of bar codes, and standardized labels. The standardization of codes and labels came about through a concerted effort by the user community. The Department of Defense led the way with its LOGMARS study; that success was followed quickly by a concerted effort by the automotive industry (the Automotive Industry Action Group). Other industries that have standardized codes and labels include the meat packing, health, and pharmaceutical industries. Others, such as the telecommunications industry, currently are involved in developing counterpart standards.

Additional developments in the control of material handling equipment include off-wire guidance of AGVs. AGV technology is one of the most prominent areas of current material handling research and implementation.[7] Functioning as a mobile robot, the AGV is being given enhanced sensor capability to allow it to function in a path-independent mode. Through the use of artificial intelligence techniques, the AGV will be able to perform more than routine transport tasks without human intervention. Using sophisticated diagnostics, it will be able to execute advanced tasks such as automatic loading and unloading of delivery trucks.

A number of European and Japanese firms are making major investments in the development of future-generation AGVs. Ranging from vehicles capable of transporting loads in excess of 200,000 pounds to those designed to transport individual printed circuit boards, a number of new entries into the U.S. market are expected within two to three years.

A related control development that will have a major impact on material handling equipment technology is interdevice communications. The manufacturing automation protocol (MAP) being developed by a number of firms led by General Motors (discussed in the section Factory Communications and Systems Technologies, under the subsection Networks) is expected to provide the common data transmission link by which many different types of manufacturing hardware, including material handling equipment, will communicate. The driving force behind this standardization effort is the desire for truly integrated manufacturing systems across the entire hierarchy of manufacturing.

Design Technology

In addition to the development of new and improved material handling equipment technology, new thinking has emerged on the design of material handling systems. Specifically, computer-based analysis, including the use of simulation and color graphics-based animation, is being used increasingly to design integrated material handling systems.

Interactive optimization and heuristics also are being applied in the design of material handling systems. Considerable research has been performed in developing performance models of a variety of equipment technologies. Trade-offs between throughput and storage capacity, optimum sequencing of storages and retrievals, and the automatic routing of a vehicle in performing a series of order-picking tasks are some of the issues that have been addressed in an attempt to gain increased understanding of material handling in the future factory environment.

DEVELOPMENTS IN MATERIAL TRANSFORMATION TECHNOLOGIES

This section describes the technologies of individual computer-controlled equipment, from numerically controlled (NC) machine tools and smart robots to computer-aided design and artificial intelligence, developments whose impetus comes from rapid advances in microelectronics and computer science. Rapid developments in very large scale integration of integrated circuits have reduced the size, cost, and support requirements of information and machine intelligence while greatly increasing its capabilities. Microelectronic technology in the future will be embedded in each machine tool and robot and at every node and juncture of computer and communication networks. The capabilities provided by this embedded intelligence will revolutionize operations on the factory floor.

Machine Tools

Although numerical control was invented and applied some 30 years ago, it continues to change the structure of machine tools in ways that still are not fully appreciated. Computerized numerical control (CNC) has replaced the punched paper tape of the original NC tools. As machines were developed specifically for NC, the traditional lines separating machine types began to blur, and two new classes of machines began to develop.

The first class, called machining centers, generally operates with a stationary workpiece and a rotating tool. Feed of the tool in relation to the work can be handled by additional movement of the tool, the work, or both. The last method is necessary for contoured surfaces and in complex cases may require more than three axes of movement, often five, and perhaps as many as eight. These machines primarily perform drilling, milling, and boring operations, but they also can tap, thread, and, when necessary, mill a surface that simulates work produced by turning.

The second new class of machines developed as a result of NC has rotating work and a stationary tool (except for feed). The machines are called turning centers and resemble lathes. They primarily do internal and external turning, drilling (of holes on the center axis), and threading, but many are equipped with powered stations that permit off-center drilling, tapping, and milling.

Both of these new classes of machines can be equipped with automatic tool-changing devices and often have automatic work loading, sensors to check on operating conditions, measuring devices, and other features that enable them to operate for long periods on different workpieces with little or no operator attention. With such versatility, a machining center and a turning center working together can perform all of the basic cutting operations on virtually any part that falls within the operating size limits of the machines.

A number of special cutting and finishing processes supplement the basic processes performed by these machines. These include gear cutting, shearing, punching, thermal cutting, grinding, honing, and lapping. Although NC was not generally applied to these operations as quickly as it was to the basic cutting operations, it is now applied to machines for each of them. (Because shearing and punching are done on presses and usually on sheet or plate material rather than on the heavier workpieces used for cutting, they are usually classified as metal-forming operations. Thermal cutting also falls in this class.)

In addition to the application of NC to traditional metal-cutting operations, several new cutting technologies have become important in many applications. The most widely used of these is electrical discharge machining (EDM), in which the workpiece is precisely eroded or cut by electric pulses jumping between an electrode and the workpiece in the presence of a dielectric fluid. Electrodes, usually made of brass or carbon, are machined to the desired form. Although the cutting process is slow, the machines operate with little or no attention, and EDM is an efficient method of cutting many types of dies.

A major recent development in EDM is the wire cut machine, in which the electrode is replaced by a fine wire sprayed with dielectric fluid. The wire slices through the workpiece as if through cheese, making shaped cuts as the workpiece table moves by NC. The wire is constantly moving between two spools so that, in effect, fresh electrode is always being used.

Low-power lasers began to be used for precision measurement about 20 years ago. Higher-power lasers are now used for welding and for sheet and plate metal cutting. Within the past two years, precision machine tools that use the laser as a cutting tool have been introduced in the United States and Japan, both for drilling and for cutting contoured surfaces. Other new technologies include the use of electron beams for drilling and welding and the use of a plasma flame for cutting.

Parts can be formed from sheet or plate in a variety of presses that bend, fold, draw, punch, and trim. The average age of presses currently in use is much higher than the average age of cutting machines, and users have generally been slower to innovate, but some press-working shops have taken advantage of new technologies. For example, some shops have installed lines in which coiled sheet is unrolled, flattened, trimmed, and shaped into parts by stamping, bending, and drawing in a continuous series of operations. Others have installed transfer presses which make finished parts from strip in a continuous series of operations. Much of the progress in forming has come through better control of the material to be formed.

The only extensive use of NC in presses has been in punch presses that combine tool-changing ability with two-axis positioning of the work for punching, nibbling, trimming, and cutting with lasers or plasma flame. However, NC controls are now beginning to appear on some other types of presses.

Tooling

Cutting tools are made from a variety of materials: high-speed steel; carbides of tungsten, titanium, and boron; oxides of aluminum and silicon (ceramics); cubic boron nitride; and synthetic and natural diamonds. Major advances in cutting-tool materials sometimes cannot be fully utilized until machines designed to take advantage of their properties are generally available.

Great progress in cutting tools has been made by applying a coating of one material (in some cases, two or three coatings of different materials) onto a base material. The proliferation of tool materials and coatings has become so complicated that computer software has been developed to aid the process. The resulting tools last longer, stay sharper, and can be used to cut hard materials such as heat-treated steel and abrasive materials such as fiberglass.

As combinations of materials and coatings produce a growing list of tooling options, the variety and volume of tooling requirements can be expected to proliferate. New product designs and performance requirements, product and process specifications, and changing lot sizes will create an ever-increasing need to match specific tooling with specific production applications. To achieve the high-quality, close-tolerance production demanded in the marketplace, manufacturers will require a large inventory of tooling to ensure that the optimal tooling is available for all production requirements. Combined with the increased expense of tooling made with rare materials and precision coatings, the costs of meeting tooling requirements will become major factors in capital budgeting decisions.

Improvements also have been made in die and mold materials. More important, however, is the change taking place in the way dies and molds are produced. Traditionally, they have been made by experienced craftsmen with a great deal of time-consuming cut and try in the finishing stages. The combination of newer EDM machines and computerized programming of die-sinking machines is removing much of the cut and try from this process.

Jigs, which serve to position the tool more accurately in relation to the work for drilling or boring, can usually be eliminated when NC machines are used. In fact, one of the major early advantages of NC was the elimination of the production and storage of jigs. Of course, if the jig also serves as a fixture to hold the work on the machine, that function is not eliminated on an NC machine.

Fixturing

Fixtures[8] hold and locate the part being worked during machining and assembly operations. The main considerations in fixture design are positioning the part in the fixture, securing the part while the machining operation takes place, positioning the fixture relative to the machine tool, positioning the cutting tool relative to the part, and minimizing set-up times. New fixturing techniques add flexibility and programmability to minimize set-up time, maximize the flexibility of the machine, and reduce storage requirements for fixtures.

The characteristics of the fixture depend on the process being performed, the shape of the part, and the tolerances required. For example, the workpiece may be subjected to strong vibrations or torque forces during some operations such as milling, while the forces in assembly operations will be much smaller. The fixtures required for these two operations are quite different and virtually incompatible. When a variety of tasks are performed, a large number of fixtures must be developed, stored, and accessed—a very expensive undertaking.

The need for a large number of fixtures remains a problem even for flexible manufacturing systems (FMSs) that can quickly and efficiently machine a number of different parts within the same part family. The FMS can help reduce economic lot sizes and reduce the expense of keeping parts in inventory. Unfortunately, this advantage is restricted by the need to have different fixtures for different parts. The cost of multiple fixtures can account for 10-20 percent of the total cost of the system, and the fixtures can sometimes cost more than the rest of the system. Clearly, the full advantages of an FMS cannot be realized without the development of flexible fixturing that can conform to different part types and machining operations.

A number of major research efforts are focused on the problem of flexible fixturing, and several solutions have been proposed. One approach would be to automate the current fixturing process, which uses blocks and clamps to align parts accurately. Instead of skilled toolmakers, robots could be used to assemble fixtures on coordinate measuring machines (CMMs). The fixtures would be mounted on standard pallets, permitting robots to load and unload parts easily and allowing easy alignment with machine tools. The CMM could cost $200,000 and vision-equipped robots at least $100,000; the hardware for the fixtures themselves and the software needed to control the robots would add to these amounts. Although the present cost may be prohibitive, this approach would ensure accurate location of the workpiece and it could be used for both machining and assembly operations.

Another approach partially encapsulates the workpiece in a low-melting-point alloy prior to machining. Encapsulation has been developed specifically for milling gas turbine and compressor blades of irregular shape. The unmachined blade is precisely positioned in the encapsulation machine. Rapidly injected molten alloy surrounds the blade and provides the clamping face, protecting the blade itself. After machining, the alloy capsule is mechanically cracked open. The problem with this approach is that the blade must be positioned accurately in the very expensive encapsulation machine, which requires a different die for each workpiece. This limits flexibility and adds expense.

A third approach is programmable conformable clamps. Developed at Carnegie-Mellon University for machining turbine blades, the clamps consist of octagonal frames hinged to accept a blade. The lower half of the clamp uses plungers, activated by air pressure, that conform to the contours of the blade. A high-strength belt is folded over the top of the blade, pressing it against the plungers, which are mechanically locked in place. Accurate alignment can be done manually or automatically with sensors. Although the clamps are limited in the types and sizes of parts they can hold and their large number of moving parts may reduce reliability, they are automatic and ensure accurate alignment.

Another approach is the fluidized-bed vise, in which small spheres are held in a container with a porous floor through which a controlled air stream passes. The spheres behave like a fluid, conforming to the contours of even irregularly shaped parts; when the air flow is stopped, the spheres come together to form a solid mass that secures the part. The advantages of this approach are that a variety of part shapes can be clamped, the clamping process is automatic, and the vise is inexpensive to build and operate. However, additional research is needed to establish a predictive model for the device and to eliminate the need for an auxiliary device to determine the location and orientation of the workpiece in the vise. Research is also under way in which electrically active or thermally active polymers are used in an authentic phase change bed instead of the pseudo phase change of the air-sphere approach.

None of these approaches offers the flexibility needed in terms of variety of applications, the types and sizes of parts that can be held, and expense. They also do not address the problem of locating the workpiece. The first three approaches use mechanical stops or surfaces, and the fourth requires an additional measuring system; this problem may be overcome by combining flexible fixturing devices with sophisticated robots.

Sensors

As machine tool automation advances, the instrumentation on the machine becomes increasingly important. Most of the early problems with automation tended to be instrumentation problems. Sensors to determine what is happening and monitoring systems to evaluate the sensor information are both needed. The role of sensors in a manufacturing environment is to gather data for adaptive control systems—for example, to supply guidance information to robots or to provide measurements for quality assurance and inspection systems. Sensors can provide automated equipment with vision, touch, and other senses, enabling the equipment to explore and analyze its surroundings and, therefore, behave more intelligently.

Sensors are currently used in factories to provide different types of data, such as the bipolar on-off of a limit switch, the simple numeric data of a temperature sensor, and the complex data provided by a vision sensor. Vision sensors, for example, can be used to determine part identification, orientation, and measurement data. Other sensors, such as tactile, acoustic, and laser range-finding sensors, are being used to measure force and shape, provide range data, and analyze the quality of welding processes.

Sensor technology is a very active field of research. Sensor research that shows promise for manufacturing includes micromechanics, three-dimensional vision for depth sensing, artificial skin for heat and touch sensing, and a variety of special-purpose sensing devices.[9] Some of the special-purpose devices have no human analog. Examples are the water vapor sensors being developed for use in sophisticated adaptive-control algorithms and the optical laser spectrometry probes that monitor chemical processes in real time. The use of adaptive closed-loop control systems in manufacturing has increased the demand for a wide variety of special-purpose sensors and has stimulated the demand for sophistication in general-purpose sensors such as vision sensors.

Other research is focused on the analysis, interpretation, and use of the data provided by sensors. Through the use of VLSI techniques in IC fabrication, intelligent sensors equipped with microchips can process data even before it leaves the sensor. For example, research is under way on vision systems that can inspect IC wafer reticles. Research on this vision system is focused on the mechanical accuracy of positioning devices, on the interface to the CAD data base describing the reticle, and on modeling the fabrication process to predict what the vision system will see. The visual information itself must be interpreted to determine whether to accept the wafer under inspection or to identify the flaw and provide feedback to correct for any imbalance. This type of intelligent sensor will eventually be integrated into many elements of manufacturing.

Model-based sensor systems such as these which use process,

CAD, simulation, and control algorithms are expected to provide manufacturing sensor systems of the future with very complex analysis capabilities. These analysis capabilities will far surpass the monitoring and control capabilities of human operators by being more sensitive, more precise in analysis, more rapid in feedback response, and more precise in corrective action. They will allow the factory of the future to work to very fine tolerances while maintaining consistently high quality control, approaching zero defects.

Smart Robots

One of the most common uses of advanced sensors is to make robots smarter. The senses of a robot are the sensors in its work cell that provide information to the robot's central controller. The "intelligence" of the robot is determined by the combined capabilities of its controller, its sensors, and its software. Most of the robots in the world's factories today have primitive controllers and software and few, if any, sensors. They mindlessly weld, paint, and pick-and-place, and some would continue to do so even if no object were present to paint, weld, or grasp. Such robots are locked into a predetermined program that does not adapt to unexpected changes in the work cell. In contrast, advanced robot systems have sensors that inform the robot of the state of its world, controllers that can interface with the advanced sensors, and software that can adapt the robot's program to reflect the changing state of its world. This is an example of adaptive behavior using a closed-loop feedback system; to a degree, it is what people do when they engage in behavior that uses the senses. It is expected that 60 percent of all robots, especially those used for inspection, assembly, and welding, will utilize vision, tactile, and other sensors within the next 10 years.[10]

Smart robots have many advantages. About one-third of the cost of a robot work cell is the fixturing that holds or feeds each part in precisely the same way each time. This cost can be saved by smart robots that can find the part they need even if it is askew, upside-down, or in a bin with other parts; it is easier to change a robot program than to change the fixturing. Smart robots will be much more adaptable to product changes because they will have less fixturing to change. Smart robots will be even more adaptable to different tasks when they can easily change their end effector for a drill, deburrer, laser, or whatever tool is required.

State-of-the-art robot systems embody elements of adaptive control and are now coming into use in factories around the world. One example is arc welding robots whose welding path is planned with the aid of a vision system that determines the location and the width of the gap to be welded. The robot software then adjusts the path and speed of the welding tool as the welding progresses. Although the welding example shows how adaptive control enables a robot to perform a task with built-in variance, the variance found in arc welding can be foreseen easily and taken into account by a human engineer or programmer. Adaptive control for robots with less-structured tasks is still in the research stage.

Robots are programmed through a special-purpose computer language. State-of-the-art languages allow the robot to perform limited decision making on its own from information obtained with its sensors. However, these programming languages are limited because they can neither interpret complex sensory data, as from a vision or tactile sensor, nor access CAD data bases to get the information they may need to identify the parts that they sense. Present languages are also robot dependent; that is, they do not allow the transfer of programs from one robot to another. This means that robots must be programmed individually by valuable, highly trained programmers.

New robot programming languages that address some of these limitations are in development in academic and commercial research laboratories. The new task level languages will allow robot programming at higher levels: the robot can be told what to accomplish or what to do with the part, and it will determine the best way to accomplish the task. The benefits expected when the new languages reach the factory floor include reducing the cost of programming, facilitating the coordination of two or more robots working cooperatively, and enabling advanced sensors to interface with the new systems.

Computer-Aided Design

Computer-aided design is not a new technology; it has already achieved wide acceptance and use in manufacturing design, and it has replaced traditional drafting techniques in other areas such as architecture and cartography. It is important to understand CAD as a technology because it interrelates with many of the other technologies described here. For example, CAD-type systems are now being used to program robots and NC machining centers. (Detailed descriptions of the interrelated roles of CAD and the CAD data base in the factory of the future are included in the sections to which they apply.)

A CAD system is composed of a graphics terminal on which can be displayed a picture of the part being designed. Designers enter the part data by drawing on a graphics tablet connected to the computer. A keyboard is used to enter dimensions and other data. The part description is then stored as one of many such part descriptions in a CAD data base. The computerized part description is not a picture, but rather a representation of coordinate points and geometric shapes from which a picture can be constructed. A particularly successful standard, the Initial Graphics Exchange Standard, has been developed for transferring data from previously incompatible representations on one CAD system to another CAD system. (This standard will be described in the Data Bases section.)

CAD offers many immediate benefits: parts can be rotated, scaled, and combined onscreen in three dimensions to enable designers to better visualize them; repetitive sections can be redrawn automatically; overlays can be easily shown onscreen; and engineering drawings can be easily updated and printed.

Other, more significant, benefits over the long run concern the use of the data in the CAD data base. These data—a computerized representation of the parts—can be used by the engineering and process planning functions, saving much reentering of data, eliminating sources of human error, and opening up a great avenue for cooperative design that includes feedback from engineering and manufacturing. Also, if the CAD data base is the only and therefore up-to-date source of part specifications, it eliminates a major current problem, concurrent use of multiple versions of part specifications.

The microelectronics industry probably has the most integrated uses of CAD. A new microchip can be designed on a CAD terminal. Once the design is in the CAD data base, the chip's performance can be simulated, the design can be modified if necessary, and the masks for the chip can be made, all automatically from the data entered at the CAD terminal. Although other industries have not yet achieved this level of integration, it has become an embodiment of the computerized design-test-modify-test-fabricate model that may change the way manufacturing entities are organized.

Graphic Simulation

As manufacturing systems come to include advanced systems such as smart robots, it becomes more and more important to be able to simulate their behavior. Two distinct kinds of simulation are now being used in manufacturing. One is the simulation, often graphic, of a single process, robot, or work cell. The second is the simulation, generally mathematical, of a system such as an FMS, a new or modified production line, or an entire factory. The former may be regarded as important in tactical or local decisions, the latter in strategic or system decisions. For this reason, graphic simulation will be treated here, and mathematical modeling will be covered later in the communications and systems section.

Graphic robot simulation is beginning to be used to select the most appropriate robot for a particular task or work cell and to plan the cell layout. The production engineer can use simulation to reject robots that do not visually appear to suit the task because of their arm configuration or timing constraints. Graphic simulation is also used for visual collision detection in the work cell, but this method is prone to error and not recommended.

Some vendors of graphic robot simulators have adapted their software to generate actual robot control programs, which is termed *off-line programming*. It permits the development of robot programs without shutting down a productive work cell, thus allowing efficient, concurrent work cell design. Although programs developed off-line currently must be used on the specific robot for which the system was designed, research to include a variety of robots in the simulation system is being conducted. For example, an off-line robot programming system has been developed that can simulate any of six commonly used robots. Researchers are also working on the related problems of how to simulate a complex sensor, such as a vision sensor, and how to debug an off-line robot program that makes decisions based on advanced sensory input. Graphic simulation and off-line programming promise to provide cost, time, and personnel savings in the efficient design of programs, work cells, and processes.

Artificial Intelligence

> Artificial intelligence is a set of advanced computer software applicable to classes of nondeterministic problems such as natural language understanding, image understanding, expert systems, knowledge acquisition and representation, heuristic search, deductive reasoning, and planning.[11]

Artificial intelligence (AI) technology will emerge as an integral part of nearly every area of manufacturing automation and decision making. Research that will affect manufacturing is being conducted in several areas of AI, including robotics, pattern recognition, deduction and problem solving, speech recognition and output, and semantic information processing. As with simulation, AI will be used at different levels in the factory of the future. Most of the AI applications will be integrated into the software that controls automated machinery, record keeping, and decision making.

Artificial intelligence is not a new field, but the maturing fruits of 20 years of AI research are just now becoming available for commercial applications. The types of AI products that will have a significant impact on manufacturing include

- expert systems in which the decision rules of human experts are captured and made available for automated decision making;

- planning, testing, and diagnostic systems; and
- ambiguity resolvers, which attempt to interpret complex, incomplete, or conflicting data.

The AI applications that deal with individual machines, processes, or work cells are described here; those that deal with system-level decision making will be integrated into the Factory Communications and Systems Technologies section.

Expert systems are in productive use today in isolated industries; petrochemical companies, for example, use expert systems for the analysis of drilling samples. Digital Equipment Corporation has used an expert system for a number of years, saving several million dollars annually in configuring the company's VAX computer systems. As human experts with years of experience become scarce, the expert system provides a way in which to capture and "clone" the human expert. An interesting feature of expert systems is that they can explain the train of reasoning that led them to each conclusion. In this way, the systems also can be used to augment human decision making, in much the same way as medical expert systems have been used. Current expert systems are best suited to situations that are somewhat deterministic when the expert's rules are known. For this reason, rapid emergence of expert systems can be expected in limited areas of technical knowledge such as chip design, arc welding, painting, machining, and surface finishing. In the 1990s, expert systems are expected that will learn from experience; this means that expert systems eventually will be developed for specialties in which there are no human experts.

Although still primarily in the laboratory, one type of AI software is attempting to simplify the use and expand the applicability of programmable equipment. For example, advanced user interfaces are now being developed that use "natural language," so that a manager can type a request at his work station in more-or-less plain English. The AI software will determine what he means, even if the request has been phrased conversationally or colloquially, and provide interactive assistance for decision making. By the year 2000, managers will probably be communicating with their work stations by voice, another application of AI techniques. Artificial intelligence technology promises to make it much easier for computers and computerized equipment to be used by personnel not having computer training, such as managers, engineers, and operators on the factory floor.

FACTORY COMMUNICATIONS AND SYSTEMS TECHNOLOGIES

In contrast to the materials and process technologies described above, communications and systems technologies tend to operate at higher levels, allowing previously separate areas of manufacturing to be integrated into systems of manufacturing. A manufacturing system is defined as a system created by the interconnection and integration of processes of manufacturing with other processes or systems. This definition implies that manufacturing systems vary from a basic system, which couples a few processes, to a hierarchical system, which integrates lower-level manufacturing subsystems into the single aggregated system. Such a system is termed a computer-integrated manufacturing (CIM) system (Figure A-1).

This variation in complexity and level makes the concept of a manufacturing system elusive to grasp. It may be helpful to think of it as an approach, a systems approach, to incrementally integrating the functions of the manufacturing corporation.

The major characteristic of manufacturing systems is their sharing of information, their communication. Traditionally, manufacturing information has been created and communicated by humans writing on paper. This paper information was based on the understanding of the human expert at that moment, although often that understanding did not accurately reflect the real state of the factory at that moment. This paper method is people intensive, time-consuming to create and distribute, often inaccurate, and in frequent need of revision. As an information communication method, it virtually guarantees delay, inaccuracy, and expense.

The advent of computer technology and network communications is changing the face of the factory floor, much as office automation has changed the front office. This technology permits

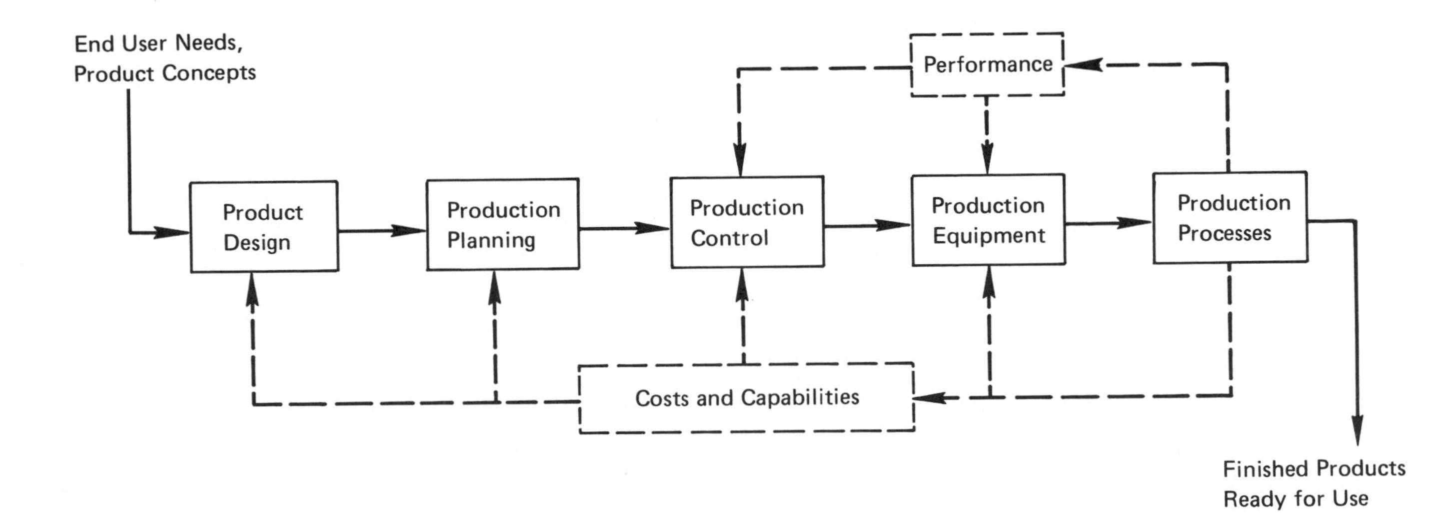

FIGURE A-1 Integration of functions in a computer-integrated manufacturing system.[12]

the system to generate its own data according to the information provided by real-time sensors built into automated machining, assembly, and inspection stations. The system gives the data to a computer, which interprets the data and takes appropriate action. This action may be to control the machining process, to replace a worn tool, or to decide whether to communicate the data, to whom, and how much.

This automated creation and sharing of information avoids the present duplication of data in several files or data bases, and it collects and communicates data at a scale and speed that will create opportunities in manufacturing never before available. For example, computer-controlled feedback permits a system to be self-diagnosing, self-maintaining, and eventually self-repairing. It allows the collection of statistical data that can be used for immediate adaptive feedback, quality control analysis, and the production of trend data. More important than any single benefit, this sharing of information makes possible the linking of systems into system aggregates. Previously disparate systems may be linked horizontally, and hierarchical adaptive control and reporting systems may be created by integrating vertically.

This information, or data, integration is the synergistic key to building manufacturing systems with broader scopes and at higher levels. The long-range goal of the manufacturing systems approach is CIM—the complete integration of the manufacturing subsystems that operate on the factory floor, the tie-in of techniques of optimization, mathematical modeling, scheduling, and data communication with the other functions (accounting, marketing, etc.) in the total manufacturing enterprise. Note that manufacturing systems are at once a means and an end.

Systems of manufacturing are integrated through the application of several technologies: communication networks, interface development, data integration, hierarchical and adaptive closed-loop control, group technology and structured analysis and design systems, factory management and control systems, modeling and optimization techniques, and flexible manufacturing systems. Artificial intelligence techniques will be embedded in, and inseparable from, most of these technologies. Communications technologies—those associated with networks, interfaces, and data bases—may be the most critical to U.S. manufacturing progress because they are the keys to the immediate development of manufacturing systems. On the other hand, technologies that analyze, manage, and optimize the system hold the greatest promise for improving the long-term competitiveness of U.S. manufacturing. These technologies will facilitate progress towards the goal of total integration from design to delivery. Each will be described in depth, both individually and as they relate to the full CIM concept.

Networks

The manufacturing network will be the backbone of factory communications and, therefore, of factory automation. Communications between tightly coupled components, such as robots and sensors, and between elements of an FMS require that data be exchanged in real time. As the complexity of the factory system increases, including the linkage from design to planning and production, the need for factory communications will continue to expand. Networks provide the physical mechanism for this communication between heterogeneous systems. The network must not only transmit the raw data but also retain its meaning, so that a different computer, running a different program, may use it. The goal of CIM is to allow all manufacturing activities using heterogeneous hardware to communicate as though they had a common language.

Networks currently provide a protocol, or agreed-upon standard, for computer communication. Most major computer vendors, as well as manufacturing equipment vendors, have defined proprietary network protocols. Thus, a variety of incompatible networks, such as Ethernet and Modway, are in use in factories today. In addition to resolving the compatibility problem, advances in network architecture are required to meet the specific communication needs of manufacturing. The speed and traffic requirements of manufacturing communication must be taken into account, as well as provisions for interfaces between otherwise incompatible networks.

General Motors and its major vendors began work on a set

of manufacturing automation protocols (MAP) for this purpose several years ago. The development of MAP has been broadened recently to include support by more than 100 major manufacturers and universities, including the National Bureau of Standards (NBS). These otherwise competitive groups realize that no single vendor can meet all the needs of a manufacturing system and that MAP may provide a solution to the communications problem between their equipment and other vendors' machines and networks.

MAP is an attempt to define the seven-level communications protocol proposed by the International Standards Organization. Although all seven levels have not yet been standardized, vendors are already selling MAP-compatible products, and farsighted purchasers are demanding that their new hardware be MAP compatible. A recent breakthrough by Industrial Networking, Inc., has put MAP on a single microchip, which will facilitate the development of factory communication networks among heterogeneous machines. While significant challenges remain, the broad membership and participation in the MAP effort can be used as a model for specifying and solving other manufacturing system problems.

Interface Standards

The network is expected to provide the physical and logical path for data communication in a factory system, but much more is required for effective communication. Networks provide the physical language and format, but do not address the semantics or effective use of the information communicated. Interface standards are needed to facilitate the effective communication of meaningful data.

The key to data integration is standardization that does not stifle innovation. Standardization of data representation within the data base is necessary to allow the full meaning of the data to be retained even when it is communicated. Current practice requires vendors of systems or modules to provide special-purpose interface definitions for each pair or family of modules that communicate. However, in some areas standards have evolved through the cooperation of users and vendors. Examples are the CLDATA file for NC machines and the Initial Graphics Exchange Standard

(IGES) for CAD data base information exchange. IGES has enabled previously incompatible CAD systems, with data stored in radically different formats, to communicate that data while preserving most of the meaning. Yet these standards rapidly grow out of date as technology moves forward. CLDATA is inadequate for nondeterministic (sensor-based) machine tool programs, and IGES does not work on solid-modeling CAD systems. The IGES continues to evolve, pointing the way to wider data integration. The challenge is to define standards that will withstand the demands of continued factory innovation or to establish mechanisms to update standards as needed.

Standards are also the solution to the interface compatibility problem that arises when equipment from different vendors is used in a network. The interface connects one machine to a communications system, which is connected to other machines, computers, and communications systems. The RS-232 interface standard is a simple protocol that has allowed communication between heterogeneous microcomputers and between computers and a host of other devices. Many machines already come with the limited RS-232 interface, but more progress is needed in standardizing manufacturing interfaces. MAP includes the definition of intelligent interfaces which can connect previously incompatible systems to a network.

The lack of interface standards can be a major impediment to achieving CIM. If well-defined information interfaces between modules or subsystems were established for the components of manufacturing systems, components could be developed independently and enlarged as advances in technology became available. This would facilitate compatibility of the equipment of multiple vendors in the heterogeneous systems expected in the factory of the future. Interface standards of this type are the basis of research at the Automated Manufacturing Research Facility at NBS. Significant questions must be answered, however, before the information interface for the modules in the manufacturing system of tomorrow can be defined.

In addition to these interface standards for information, two other kinds of interface standards are needed. The first and most neglected is the interface between human and programmable systems. The second is the physical interface between mechanical systems.

The man-machine interface includes the commands to be given by the human to make the machine perform a task successfully and the input device or physical method—keyboard, joystick, light pen, or voice—for entering those commands. Most current programmable systems are commanded through a programming language that is proprietary to the vendor of the system. This has given rise to a Tower of Babel of control languages requiring highly trained programmers to control modern manufacturing systems. No programmer can begin to master all of the languages and input devices found in an automated factory.

Two recent trends are expected to ease the interface problem between nonprogrammers, such as engineers and technicians, and the increasingly complex programmable automation systems found in the areas of robotics, NC tools, material handling, and processing systems. Hierarchies of languages and personnel are being developed in which highly trained programmers will deal with the raw control languages and sophisticated control algorithms, less-skilled programmers will deal with a higher-level simplified language, and equipment operators will not use actual programming languages at all. This hierarchy—automation systems programmer, applications programmer, user programmer, and user—parallels the evolution of personnel in the computer field.

The second trend is the use of AI to develop task-level control languages (discussed above in the Smart Robots section). Currently under research, task programming systems will reduce programming requirements to the steps on a common process planning sheet so that programmable manufacturing systems of the future will be controlled by statements similar to those one would give to a person doing the same task. These advances will provide new generations of specialized, user-friendly manufacturing subsystems that will make the most of factory personnel.

The mechanical interface problem for the factory is solved most easily by the development and adoption of standards. The lack of standards for the newer systems is a major impediment to progress. Examples of mechanical interfaces in need of standardization include

- robot end-of-arm and gripper attachments;
- pallets, totes, and other part conveyances;
- the mechanical interface for the loading and unloading of parts and pallets at machining centers; and
- the interface between robot carts and material handling systems.

Progress with the mechanical interface problem requires the usual consortium to agree on and promulgate standards. The major roadblock has been the lack of an organized body, leadership, and focus on the problem.

Data Bases

Network and information interface standards are the means of sharing data, but it is not enough to move data from one application to another. The data must be stored, and the semantics, or full meaning, of the data must remain intact when the data are retrieved. A data base provides the long-term memory, or storage facility, that contains the manufacturing data, and the information retrieval system extracts specific data from the data base.

Current practice finds a large number of information retrieval systems in place even at the same company. Each system is associated with one major function, such as accounting, shop floor information, material requirements planning, or quality control statistics. Many of these data systems are state-of-the-art information retrieval systems with complex functions to maintain and update the data base. Unfortunately, the different data bases contain redundant and conflicting data in incompatible formats. Furthermore, many of these systems are dedicated to a single computer and use proprietary data representations that are incompatible with those of other systems. Thus, the task of developing an integrated manufacturing data base management system that can include every major function in a factory is formidable.

The most serious immediate barrier to the integration of manufacturing data is the incompatibility of CAD data with information needed by computer-aided engineering (CAE) and process planning. CAD technology has become highly effective in capturing the geometry of parts, including the description of dimensions, shapes, and surfaces. Real parts, however, are made up of smaller components and may themselves be components of a larger assembly. The CAD data base currently cannot capture the relationship of the parts to the whole, but both CAE and process planning require detailed attention to the joining of separate parts, their mating surfaces and tolerances, and their overall dimensions after assembly. The CAD data base does not include knowledge or specification of materials, but CAE needs material data for its engineering analyses, process planning needs it in the creation of NC programs, and material handling needs it to select material from inventory.

A second serious barrier to the integration of manufacturing data is the current inability to model the processing portion of the overall system (the two right-hand boxes of Figure A-1). Such a model would allow information on production costs and capabilities to be fed back, on-line, to the product design activity as it is being performed. This capability is essential to optimizing the producibility of products at the design stage. With this capability, each decision proposed in the engineering design process would result in simultaneous information on the effect of that decision on production costs and required capabilities (relative to available capabilities) for production of the product. It also would result in major cost savings in the production activity, since it is well known that the majority of production costs are frozen at the engineering design stage. While such computer-based integration of manufacturing data is technologically feasible, many difficult problems must be solved to bring it into being.

A further problem with the data in a CAD data base is that the geometrical data cannot be searched or aggregated in the ways that have become standard for textual data. Without explicit hand coding, it is not possible, for example, to retrieve all parts which use a particular fastener. Group technology (discussed later in this section) is an attempt to code and classify the geometry, function, and process data in a way that will permit the use of standard retrieval functions.

One of the keys to the data integration problem lies in the development of flexible data schemas. A schema is a method of storing data so that its meaning and accessibility are retained. Most data bases use rather fixed data schemas that restrict the new types of information that may be added and limit data retrieval capabilities. Future data bases will have more flexible schemas so that, for example, materials information can be added to the CAD data base by an engineer at a CAE station or by an expert system that contains knowledge of materials and applications.

Beyond the compatibility problem are other technical challenges to the implementation of manufacturing data base systems. For example, experts predict that future manufacturing data bases will be 20-50 times larger than present data bases. The size of the data base, the time tolerances for communication, and the variety of users suggest that a manufacturing system data base will be distributed across multiple heterogeneous systems, which may be in different geographical locations. This presents significant technical challenges to the achievement of a logically integrated manufacturing data base. The concepts and protocols normally used to ensure proper access, control, and update will need to be expanded to meet this sophisticated method of data base organization. Interim solutions in place today are neither geographically nor heterogeneously distributed, but progress toward these goals is being made.

A last challenge posed by the manufacturing data base is the use of probabilistic or incomplete data. Current data base systems can only represent facts and cannot deal with uncertainty or conflict within their data. Manufacturing information systems of the future will depend on AI to deal intelligently with this type of information.

One of the most critical roles of people in the factory of the future will be to interact with intelligent manufacturing systems through work stations, terminals, or networked microcomputers. As expert systems and other forms of AI become embedded in systems of manufacturing, the systems will be able to perform more and more of the decision-making tasks previously performed by people. At first, these automated decisions often will have to be reviewed by people and then interactively modified, much as an architectural plan takes shape in a dialogue between client and architect. People without knowledge of the data schema will routinely query the system for information needed to make decisions. The data retrieval system will have to determine exactly what is important to the inquirer and then retrieve and massage the appropriate data. The person may even want the system's "opinion," or the system may ask for the person's opinion. The factory of the future will regard personnel and intelligent systems as partners in a dialogue that should encourage very sound decision making.

Group Technology

Group technology[13] (GT) is a key philosophy in the planning and development of integrated systems. In practice, GT is defined as a disciplined approach to identifying by their attributes things such as parts, processes, equipment, tools, people, and customer needs. These attributes are then analyzed to identify similarities between and among things; the things are grouped into families according to similarities; and these similarities are used to increase the efficiency and effectiveness of managing the manufacturing process.

Although it is relatively simple to define GT, it is difficult to create and install a GT system because of the difficulty in defining clearly how similar one part is to another. For example, parts can be categorized in terms of shape or manufacturing process requirements. These two different viewpoints require a flexible approach to the GT data base and the realization that parochial, departmental views of coding may allow some localized cost saving but miss the large corporate savings possible.

The GT concept requires that the attributes of a thing, such as a part, be identified and classified. Attributes can be visual, such as the surface finish or shape of a part; mechanical, such as the strength of the material; or functional, such as the clock aspect of a printed circuit board. The attributes may also be related to the environment of the part, such as the processes or equipment necessary to make it. Because of the many possible coding strategies, it is hard to know in advance exactly what attributes will be important as the GT data base is used by more and more kinds of software. It is therefore important to guarantee that the data base structure is flexible enough to add attributes and to modify coding schemas as necessary for new applications.

The four basic GT applications are design retrieval, FMSs, purchasing support, and service depot streamlining. Design retrieval is a GT application in the design engineering area to provide the maximum potential for part standardization. Moreover, it permits greater cooperation between the design engineer and manufacturing engineer by providing feedback about specific part attributes in the GT data base. Design retrieval will also supplement product reliability data based on the actual performance of parts with similar attributes. Finally, by determining the relationships of new parts to previously designed parts, it increases human productivity in the generation of new designs and the revision of old designs. The software necessary to implement design retrieval is the simplest of all GT software; it involves a simple query to a GT data base for specific feasible ranges of variables.

Flexible manufacturing systems exemplify a more sophisticated and more profitable GT implementation. Such an FMS can be created by identifying a cluster of machinery that can, or will be able to, service a particular family of parts. The FMS can then be streamlined to produce this part family optimally.

Purchasing support is a rather new GT application, yet almost every major manufacturer has a quasi-GT system, called a commodity code, already performing this task. A rigorous GT system may pay a tremendous financial reward by permitting all related parts, including those that do not obviously belong to the same families, to be identified throughout a factory or corporation. In one instance, a vendor of hoses offered a 50 percent reduction on hose prices if a corporation could identify all hoses and their attributes to be purchased over a given period. The corporation saved millions of dollars by rigorously following a GT system to identify all hoses to be ordered. A GT purchasing support system offers buyers a significant way to cut costs through knowledge and buyer leverage.

Service depot streamlining is a GT application which can help determine the most advantageous service parts strategy by identifying where alternate parts may be used. Standardization of parts in the service depot allows for substantial reduction in inventory and repair time, even if the standard is the most expensive item.

The premise of a GT system is that the similarity of parts and processes can be turned into substantial cost savings. The most far-reaching applications of GT will be made possible by the structuring of the parts data base itself. If the data base information is captured in attribute form and linked to applications by similarity, the ability of the data base to support manufacturing decision making will be greatly enhanced. This global application of GT to data base design is only now gaining popularity, and research on it is still in its infancy.

One of these far-reaching applications is in the area of process planning. By performing a rigorous analysis of manufacturing processes and parts to be made, a manufacturer will improve his ability to move from the present method of process planning into the more highly integrated future. For example, a manufacturer may evolve from his present variant process planner, which uses GT to match part families to process families, to a more sophisticated generative system in which more knowledge captured in the GT data base will be used to optimize the process plan. Eventually, a highly integrated system can be achieved that delays the final process planning step until the part is to be made, optimizing not only the process but also capacity utilization. Group technology, like network technology, will be a cornerstone of CIM systems.

Adaptive Closed-Loop Control

One difference between an automated system and an intelligent system is the amount and kind of feedback that is generated from an activity and passed up to a decision-making entity. This feedback allows a system to know its own state, to know when it is out of balance, and to respond to the imbalance until stability is achieved. This adaptive closed-loop control will be used at all levels of manufacturing systems from sensor-based feedback to robots or NC tools, to inspection station-based feedback to a cell controller, to a factory floor data collection system that feeds back to process planning and scheduling systems. This property of adaptive feedback is the key to improving product quality, with zero defects a realistic goal (see Figure A-1).

Adaptive feedback is also one key to better management, since only in this way can a manager know exactly the state of production, including exact costs. Systems of manufacturing have a feedforward property that will allow management to control the factory floor with an effectiveness and immediacy never before possible. (A more global discussion of hierarchical control is found in the subsection Computer-Integrated Manufacturing Systems.)

Feedback and feedforward properties can also provide machines and systems with self-diagnosis features. Thus, a manufacturing system can tell if something is wrong with it and what is wrong and can suggest the remedy to a higher entity. For example, researchers at the AMRF have already demonstrated the ability to sense when a tool is about to break so that automated equipment can change the tool without the disruption caused by untimely failure. Next will be limited self-maintenance and repair capabilities. When a data-intensive system breaks down, the integrity of that data is threatened. Self-diagnosis will inform the data base system of the integrity of the data and, if it is threatened, the system will take either conservative or remedial action.

Factory Management and Control

The factory of the future will be managed and controlled through automated process planning, scheduling, modeling, and optimization systems. The successful implementation of large-scale factory-level systems depends upon structured analysis and design systems that depend heavily on GT. Limited structured analysis systems, such as the Air Force-sponsored Integrated Computer-Aided Manufacturing Definition, have been in use for years in the analysis and design of large projects. Only through such systems can a manager know the exact state of his factory, and only through such exact knowledge of the present can a manager intelligently implement systems of manufacturing for the future. New systems development methodology packages, such as STRADIS, promise help in this area, but much work remains to be done before such systems are easily used by the actual decision makers.

Similar work must be done on process planning and scheduling systems before they can use the feedback and feedforward properties of the hierarchical and adaptive closed-loop control systems to be found in future manufacturing systems.

Management functions will be hierarchically distributed so that "go" and "halt" decisions may be made effectively from many levels and by human or machine. Through the use of terminals on the factory floor and throughout the decision-making structure, the system can respond instantaneously to human command. At first, most of the decision making will rest in the hands of humans. Low-level manufacturing systems now work in this way. As the systems become integrated at higher and higher levels, decision rules and methods will be built into them; systems will develop plans to carry out human-specified activities. On the authorized human's approval, the system will carry out the task, making low-level decisions on its own. If a low-level decision-making entity does not have a certain level of confidence in its decision, it may pass the decision up to the next-higher entity, be it human or computer. This new kind of man-machine interaction will allow humans to do what they do best: create, define, and communicate. The machine will do what it does best: work hard, steadily, and accurately.

Modeling and Optimization Systems

One tried and true method of representing an activity to a computer is through mathematical modeling. Computerized modeling tools, such as SLAM, have been used by simulation experts for years, but simulation is still more an art than a science. With experience and feedback, our ability to represent complex activities mathematically will be refined. A modeling and simulation package will be a necessary part of an intelligent structured analysis and design system. It is hard to overstate the importance of structured analysis and design systems; they will operate at high levels, with much built-in decision making.

System modeling will become a commonplace and necessary prerequisite to the successful design and implementation of large-scale manufacturing systems. This is because large projects contain too many facets to be managed effectively with only human memory and computation capability. Artificial intelligence techniques will be needed to reduce the tremendous amounts of data generated by such systems to humanly understandable terms. This intelligence must be of a higher order than the AI expert systems in existence today.

With the addition of AI, a modeling system can become an optimization system, guiding its human managers to the most productive, most cost-effective, or highest-quality utilization of resources. With such optimization capability, managers will be able to sit at their work stations and, in real time, analyze the various possibilities to determine optimal solutions and mixes. The availability of accurate information on cost, time, and quality will eliminate much of the guesswork in manufacturing decision making.

Flexible Manufacturing Systems

Flexible manufacturing systems are expected to dominate the factory automation movement within 10 years. These FMSs will be tied into larger-scale manufacturing systems, but it is valuable to consider the FMS as a critical unit or building block in total factory integration. An FMS may be described as an integrated system of machines, equipment, and work and tool transport apparatus, using adaptive closed-loop control and a common computer architecture to manufacture parts randomly from a select family. The hardware components of an FMS may include an NC tool, a robot, or an inspection station. The part family processed by the FMS is defined by GT classification. For greatest productivity, the FMS is optimized to produce only one family of parts, and conversely, the parts produced by the FMS are designed to facilitate processing by the FMS.

The concept of flexibility as used in an FMS includes

- use of GT to achieve a part mix of related but different parts;
- batching, adding, and deleting of parts during operation;
- dynamic routing of parts to machines;

- rapid response to design changes;
- making production volume sensitive to immediate demand; and
- dynamic reallocation of production resources in case of breakdown or bottleneck.

Flexible manufacturing has been a reality in U.S. industry since its introduction in 1972, and the number of new FMS installations is doubling every two years. The number and flexibility of FMSs is expected to increase, and the cost of FMS installations is expected to drop. Although the United States was largely responsible for the technological development of the FMS, Western Europe and Japan both have more FMS installations than this country. In fact, one of the most frequently cited FMS installations is located in the Messerschmidt-Boelkow-Blohm (MBB) plant in Augsburg, West Germany. The basic elements of this FMS are 25 NC machining centers and multispindle gantry and traveling-column machines; automated tool transport and tool-changing systems; an AGV workpiece transfer system; and hierarchical computer control of all these elements. The FMS is used to build wing-carrythrough boxes for Tornado fighter-bombers. Comparisons by MBB of the performance of this FMS versus the projected performance of stand-alone NC machine tools doing the same work clearly show the advantages of the FMS approach:

- number of machine tools decreased 52.6 percent;
- workforce reduced 52.6 percent;
- tooling costs reduced 30 percent;
- throughput increased 25 percent;
- capital investment 10 percent less than for stand-alone equipment; and
- annual costs decreased 24 percent.[14]

U.S. statistics for FMS installations are no less startling. An FMS at General Electric (GE) improves motor frame productivity 240 percent; an AVCO FMS enables 15 machines to do the work of 65; and at Mack Trucks, an FMS permits 5 people to do what 20 did before. In addition to productivity enhancements, the FMS offers increased floor space capacity. GE, for example, reported that floor space capacity was increased 50 percent, with a net floor space reduction of 30 percent. An FMS can make a factory more responsive to its market—GE reported a shortening of its manufacturing cycle from 16 days to 16 hours.

Computer-Integrated Manufacturing Systems

The technologies discussed above will be integrated to create CIM systems whose synergy will make the whole greater than the sum of its separate technologies. CIM systems promise dramatic improvements in productivity, cost, quality, and cycle time. However, since full CIM has not yet been accomplished and depends on continued technological progress, the benefits are difficult to quantify accurately. Incremental gains from the implementation of individual technologies and subsystems will be substantial. These benefits are illustrated by the following data from five companies that have implemented advanced manufacturing technologies over the past 10-20 years:[15]

Reduction in engineering design cost	15-30 percent
Reduction in overall lead-time	30-60 percent
Increase in product quality	2-5 times
Increase in capability of engineers	3-35 times
Increase in productivity of production operations	40-70 percent
Increase in productivity of capital equipment	2-3 times
Reduction in work-in-process	30-60 percent
Reduction in personnel costs	5-20 percent

The cumulative gains of total system integration can be expected to build on these results exponentially.

The long-range goal of CIM is the complete integration of all the elements of the manufacturing subsystems, starting with the conception and modeling of products and ending with shipment and servicing. It includes the tie-in with activities such as optimization, mathematical modeling, and scheduling.

A CIM system is created by the interconnection or integration of the processes of manufacturing with other processes or systems. The resultant aggregate system provides one or more of the following functions or characteristics:

- An information communication utility that accesses data from the constituent parts of the system and serves as an information communication and retrieval system.
- An information-sharing utility that integrates data across system elements into a unified data base.
- An analysis utility that provides a mathematical model of a real or hypothetical manufacturing system. Employing simulation and, when possible, optimization, this utility is used to characterize the behavior of the modeled system in various configurations.
- A resource-sharing utility that employs mathematical or heuristic algorithms to plan and control the allocation of a set of resources to meet a demand profile.
- A higher-order entity that integrates information and processing functions into a more capable, effective processing system.

These functions and characteristics are not mutually exclusive in actual manufacturing systems; rather, they overlap significantly with all of the elements interconnected and integrated continually to form a single aggregated CIM. Perhaps the most important and least understood step in this process is the creation of an integrated system that is a higher-order entity; this is the true system-building goal.

Both horizontal and vertical growth of CIM systems can be expected as the year 2000 approaches. State-of-the art technology now includes small aggregates of computer-integrated tasks, often called islands of automation. Such islands of automation are found in design, where CAD work stations from different vendors share their data through a common data base and data conversion interfaces; in planning, with manufacturing resource planning (MRP) systems; and in production, where a work cell composed of a robot, machine tool, and inspection station may be coordinated by a cell controller.

In leading-edge plants, several of these islands of automation have been aggregated into larger manufacturing subsystems, termed continents of automation. At this level of integration, links exist between the design and engineering departments, with CAD terminals and data bases sharing data with CAE work stations and data bases. In planning, MRP can be linked to traditional data bases containing ordering and shipping information. On the factory floor, several work cells may be integrated with a material handling system to create an FMS.

In the factory of the future, these continents of automation will be integrated into worlds of automation that will eventually encompass not only entire factories, but also entire corporations. Because of the volume of data and complexity of decisions needed for full integration, a hierarchical structure is the only feasible way to achieve it.

A hierarchical structure has certain implications for the architecture of CIM systems. Data use and decision making must occur at the lowest levels possible. Only certain summary data will be passed upward in the hierarchy to be used in reporting the factory's state and in statistical trend analysis. Thus, an information and decision hierarchy is needed that practices management by exception. If additional information is required at upper levels, it will be requested. If local decision making cannot resolve a conflict, a decision will be requested from above. Conversely, management decisions may be communicated almost instantly throughout the system for rapid compliance.

The hierarchical structure further implies the use of distributed data bases and distributed processing. A mainframe computer may be the host computer to the factory of the future. Connected to it will be an array of minicomputers, one level down in the hierarchy, each acting as host controller to an intermediate-level manufacturing system. A mix of local and centralized data storage will be appropriate for each computer. Below the minicomputers will be microcomputers acting as cell controllers, graphics work stations, or executive work stations.

The elements of this hierarchical structure can be thought of as subsystems, categorized by the role they play, although it must be remembered that categories may overlap considerably. Most subsystems of manufacturing fall into one of the following broad categories: (1) information and communication, (2) integration of processes, or (3) resource allocation. Note that each category cuts across traditional manufacturing boundaries. An information-communication subsystem, for example, could include a network that permits the geometric part data stored in the CAD data base to be transformed through GT techniques into an actual process plan and then into robot and NC programs communicated to the factory floor. The data would then be transformed and communicated for process scheduling and material handling, right up to the delivery of the finished product.

Information-oriented subsystems include traditional management information system and data processing roles. These subsystems will be able to expand to include geometrical data from CAD systems, material and process data from GT coding, parts-in-process data, and order and inventory data. Information subsystems will have analytic capabilities by which the data can be massaged for quality control and trend analysis. Data retrieval will be easier for operators and decision makers through the use of new query languages or programs that will allow nonexperts access to complex data. Most, if not all, manufacturing subsystems have strong information and communication functions, even if they are primarily process or resource oriented.

Manufacturing subsystems on the factory floor will integrate traditional manufacturing processes by coupling and controlling previously separated processes and by carrying out computer-generated process plans. At the lowest level, this will involve data communication from sensors to a computer-controlled machine or robot. This provides the real-time adaptive control necessary to improve the work quality and throughput of individual stations. At the next level, factory floor manufacturing subsystems can integrate several processes, such as an NC machining station, an automated inspection station, and the robot which services them. In this example, the coordination is supplied by a computer which controls the work cell. The process plan is downloaded from a computer, which may be in the CAE area, to the work cell controller, which coordinates the processing by the machines in its cell. With automated inspection and data collection, the process plan may be modified to eliminate defects by responding in real time to tolerance changes. At yet a higher level, work cells are integrated into an FMS so that an automated scheduling system can assign a part in process to the next available work cell that can perform the necessary operation. This allows a system with fewer parts in process, shorter queues, fewer holding areas, and much more efficient use of floor space.

Resource allocation subsystems span a broad scope from small-scale material handling systems serving individual work cells to broadly implemented systems that monitor and control inventory, schedule work, and allocate materials to the factory floor on tight schedules. Automated material handling systems can be integrated into work cells and families of work cells to produce a powerful FMS. In turn, the FMS can be linked to production planning and capacity planning systems to form the fully computer-integrated manufacturing systems illustrated in Figure A-1.

One of the most important reasons for implementing small-scale subsystems of manufacturing now is that they can be successively integrated into these higher order entities, CIM systems, that benefit from the synergy between operational programs, product data, and process data.

CONCLUSION

All of these advanced manufacturing technologies, from materials and machine tools to the subsystems and CIM systems, provide the ability to perform traditional manufacturing tasks in a highly advantageous but nontraditional manner. Many of the individual technologies and subsystems of manufacturing can be implemented today and, in fact, must be implemented soon for a manufacturer to remain competitive. Real progress toward the factory of the future will take place through the higher-level integration of these technologies. Although a handful of domestic manufacturers continue to make progress in implementing and integrating many of the technologies described in this appendix, real barriers to full integration remain.

Specifically, standards are critically needed for the definition and communication of part data. At higher levels, the need is for proven systems of hierarchical control and feedback and usable methods of automated classification of parts and processes (GT). Required at the highest level are the evolution of structured analysis and design systems that include modeling and optimization packages, as well as intelligent user interfaces that can be used interactively by managers in real time. The technology is here now or just around the corner. U.S. manufacturing needs far-sighted management and trained manufacturing engineers to put the pieces together.

NOTES

[1]More detailed and comprehensive information on current developments in metal matrix composites can be obtained from the National Research Council's National Materials Advisory Board.

[2]The New York Times. November 17, 1985. Goodbye to Heavy Metal, p. F-1.

[3]This section is based on Research Briefing Panel on Ceramics and Ceramic Composites. 1985. Research Briefings 1985, pp. 60-71. Washington, D.C.: National Academy Press.

[4]Advanced Ceramics Technology. Pp. 23-24 in Technology Today, March 1986.

[5]Committee on the Status of High-Technology Ceramics in Japan. 1984. High-Technology Ceramics in Japan, p. 30. Washington, D.C.: National Academy Press.

[6]Ibid, pp. 24-25 and 33-34. Gives market growth projections in Japan, as well as different applications given several price scenarios.

[7]Interestingly, the initial patents for AGVs were held by a U.S. firm, the Barrett Corporation. However, consistent with the European dominance in material handling equipment technology that has existed for many years, the Barrett Corporation was acquired by a European firm, the Mannesmann Demag Corporation.

[8]This section is based on Thompson, Brian, 1985, Fixturing: The Next Frontier in the Evolution of Flexible Manufacturing Cells, CIM (March/April):10-13.

[9]A good description of current developments in micromechanics can be found in Brandt, Richard, 1986, Micromechanics: The Eyes and Ears of Tomorrow's Computers, Business Week (March 17):88-89.

[10]Smith, Donald N., and Peter Heyler, Jr. 1985. U.S. Industrial Robot Forecast and Trends: A Second Edition Delphi Study. Dearborn, Mich.: Society of Manufacturing Engineers.

[11]Committee on Army Robotics and Artificial Intelligence. 1983. Applications of Robotics and Artificial Intelligence to Reduce Risk and Improve Effectiveness, p. 58. Washington, D.C.: National Academy Press.

[12]Committee on the CAD/CAM Interface. 1984. Computer Integration of Engineering Design and Production: A National Opportunity, p. 11. Washington, D.C.: National Academy Press.

[13]Shunk, Dan. Integrated Cellular Manufacturing. Paper presented at the WESTEC conference, Los Angeles, March 1983.

[14]Computer Integration of Engineering Design and Production, p. 50.

[15]Ibid., p. 17.

APPENDIX B
Management Accounting in the Future Manufacturing Environment

The new manufacturing environment will require a major rethinking of corporations' management accounting systems. Computer-integrated manufacturing (CIM) equipment requires extensive digital data for command and control of operations. These same data can be used to measure resource utilization and the quantity and quality of output on a continuous basis. For example, the time each product spends waiting to be processed and actually being processed on each machine is readily available. Also, physical counts on actual output, rework, and scrap can be tracked continuously. Therefore, it becomes possible to compute actual costs on a real-time basis by individual product classes.

In contrast, traditional cost measurement systems compute costs and output over fairly long periods, typically a month. Because of the time required to collect this information, the monthly cost report is not available until the middle of the subsequent month. Also, the data are aggregated, so monthly variances are difficult to trace to individual products or batches.

The new manufacturing technology provides an opportunity to move from cost measurement on a delayed, aggregate basis to performance measurement on a continuous, hourly, or daily basis, by resource category, cost center, and product class. Such new measurements will permit better understanding of product costs and permit more rapid detection of problems and opportunities for learning and improvement of manufacturing processes.

The CIM environment will also make cost accounting systems based on direct labor obsolete. Many companies still distribute costs from the factory and from cost centers to products based on direct labor burden rates. Such a system may have been reasonable decades ago, when these systems were designed, because direct labor was the most expensive component of total manufacturing cost. In highly automated environments, however, direct labor is a very small percentage of total costs, 5 percent or even much less. The direct labor that is used is more concerned with set-up and supervision than with actual processing of output. As overhead and indirect costs have escalated and direct labor costs have diminished, the burden rate on direct labor has reached 500 percent or, in highly automated environments, 1,000 to 2,000 percent. Clearly, this is a confusing way to trace factory costs to products.

More imagination is needed for tracing costs to products, which inevitably will require new and multiple overhead allocation bases. For example, costs associated with materials (purchasing, traffic, receiving, distribution, and storage) can be traced to materials purchases based on material dollars or on quantity, size, or weight of materials, as seems reasonable. Costs associated with acquisition, maintenance, repair, and operation of machines can be traced to products on a machine-hour basis. Costs of production control and expediting departments can be traced to products based on the number of set-ups. The costs of engineering, quality assurance, and customer support and service should be traced to the products which require or which benefit from the use of these departments. Since materials, equipment, and overhead will be the most important manufacturing costs, it becomes imperative to develop cost accounting systems that will trace these costs to products rather than rely on arbitrary allocations based on direct labor, a diminishing component of manufacturing costs.

The present monthly cost accounting reporting cycle, with its allocation of overhead based on direct labor content, exists mainly to distribute manufacturing costs between goods sold and inventory so that monthly profit and loss statements can be computed. This financial reporting objective has been the main driver for cost accounting systems. It has created the situation in which cost data are too late to be very useful in cost control and too aggregated to be helpful in identifying actual costs for product mix and pricing decisions.

This financial accounting objective of measuring product costs and profits for periods as short as a month or a quarter, or perhaps even a year, will be undermined by the new manufacturing technology. Most manufacturing costs will become fixed costs—product design, equipment and facilities, salaries and wages—which will not vary with the volume of production. Indeed, many of these costs must be incurred before any items are produced; they are sunk costs. The primary variable costs of production will be materials, some energy to operate equipment, and maintenance and repair of machines. This type of cost structure implies very high gross margins; that is, variable costs must be a small fraction of selling prices if the firm is to recover its investment in product and process development, equipment, software development, and prototypes and testing.

With this pattern of cash expenditures and cash inflows, income measurement over short periods becomes a meaningless exercise since it is wholly dependent on arbitrary assumptions about the amortization of prior-period expenditures to current production and sales. The more meaningful measurement of income occurs over the full span of the product's life cycle. By how much will the net cash revenues (sales less variable costs and period expenses) exceed the investment required to get the product into production? At what rate are current and expected future sales generating cash to recover the cash invested in product and process development? At the end of the product's life, management will be able to know whether and by how much the product generated cash in excess of that invested in it, but to apportion this income or loss to arbitrarily short periods within this life cycle will not be an interesting or relevant exercise.

However, saying that income cannot be measured over short periods does not imply that meaningful short-term performance indicators cannot be devised. Certainly the rate of cash recovery will be an important financial measure, but a variety of nonfinancial measures will be equally important. These should include key indicators of manufacturing, marketing, and research and development success. For example, measures of quality, including internal and external failure rates, yields, and rework, can be computed. Measures of production efficiency, such as machine availability, throughput and lead times, average inventory levels, and set-up times, can be calculated. Percentage of delivery times met, whether product development milestones are being achieved, customer satisfaction measures, employee absenteeism, turnover, skill levels, and morale will be much more interesting short-term performance measures than a financial profit figure requiring arbitrary allocations of expenditures on past, current, and future product and process developments. Thus, the new manufacturing environment not only will require entirely new cost measurement and management systems, but also will cause managers to abandon attempts to measure profitability over periods as short as a month, quarter, or even a year. Nonfinancial measures may provide a better indicator of a firm's ability to earn long-term profits than any short-term profit figure managers may attempt to compute.

Finally, firms will find it necessary to expand their procedures for justifying investment in new process technology. Current procedures tend to emphasize easily quantified savings in labor, materials, or energy. Further, these savings tend to be recognized for arbitrarily truncated periods, sometimes only one or two years. In addition, new equipment is justified on the assumption that the firm will continue to do the same volume of business even if the new investment is not undertaken. None of these assumptions will be valid or helpful when contemplating investment in new process technology, for several reasons.

First, while CIM equipment offers significant direct labor savings, it also offers considerable improvements in quality, inventory and floor space reduction, great reductions in throughput and lead times, and flexibility to accommodate product redesigns and new generations of products. To consider only easily measurable labor savings significantly understates the benefits of CIM technology.

Second, the benefits from CIM technology will persist over long periods. It is unlikely that the hardware and software investments in CIM technology can be repaid in a couple of years, but the flexibility of this technology ensures that its useful economic life will be much longer than that of conventional dedicated equipment. Therefore, analytic capital authorization procedures must use realistic estimates of the considerable useful economic life of this equipment.

Third, firms that do not invest in the new technologies will soon find that they are no longer the low-cost, high-quality producers and will experience declining net cash flows. Therefore, the correct way to evaluate new process technology is not to extrapolate from past and current cash flows, but to find a pattern that predicts the decline in cash flows owing to anticipated adoption of new technology by competitors.

In summary, the new manufacturing technology will require major modifications in the way firms measure and manage costs, in their measures of performance—both financial and nonfinancial—during short periods, and in the way they justify investments in new technology. Failure to make these modifications will inhibit the ability of firms to be effective and efficient global competitors in the manufacturing environment of the twenty-first century.

APPENDIX C
A Review of Policy Recommendations of Selected Study Committees, Panels, and Commissions, 1979-1985

INTRODUCTION

The past decade has seen growing concern about signs of a weakening of the position of U.S. manufacturing in the world economy. Numerous committees, commissions, and panels have been appointed to study the sources of retardation and to recommend policies to strengthen U.S. industry's competitiveness. While some of these study groups have dealt with broad issues of productivity, innovation, and technological change, the main focus has been on the manufacturing sector.

This appendix summarizes policy recommendations made by 17 study committees during 1979-1985. The sponsors included the Committee for Economic Development, the American Productivity Center, the National Research Council, the Office of Technology Assessment, and the U.S. Department of Commerce. The recommendations reflect a consensus among business executives, university experts, and government officials. In some cases, union officials and public representatives also participated. Each study group is briefly described, and its major policy recommendations or options are outlined.

OVERVIEW OF MAJOR POLICY RECOMMENDATIONS

To provide an overview of the scores of policy recommendations included in these various studies, they have been classified and briefly analyzed under 10 headings.

Research and Development in Science and Technology

Continuation or increase of the federal government's support of basic research was favored by several study groups. Support for generic research, particularly in manufacturing technology and automation, also was favored, as was coupling industry and university research. The President's Commission on Industrial Competitiveness recommended the establishment of a federal Department of Science and Technology to encompass all existing programs.

Human Resource Management

Various studies included wide endorsement of the idea that employees should be more involved in the decisions affecting their work, that the adverse human impacts of automation need to be moderated, that reward systems linked to productivity performance can be helpful, and that labor-management cooperation should be encouraged.

Education and Training

Several panels stressed the need to improve and expand engineering education in a time of explosive growth in technology. Particularly vital is the need for graduate education to fill faculty vacancies at the university level. The spread of computers also is expanding the training requirements of managers and employees. According to one panel, effective computer education will require further research on educational software.

Regulatory Reform

The use of performance standards, cost-benefit analysis, and economic incentives was recommended by many panels as a way of balancing regulatory and business goals. Several panels favored a faster review process in their industries.

Tax Policy Changes

More favorable treatment of research and development (R&D) in the tax code was recommended by many groups, along with continuation of incentives for capital investment. Several panels favored simplification of the tax structure.

Capital Investment

The capital investment issue was considered by several groups as connected with efforts to stabilize monetary and fiscal policy. According to the President's Commission on Industrial Competitiveness, the supply of capital should be increased by reducing the federal deficit.

Trade Policy

Many committees favored closer monitoring and stricter enforcement of import policies. A few favored a review of the effect of national security export controls on competitiveness. Two panels endorsed the establishment of a single agency or Department of Trade.

Antitrust Policy

Modification of antitrust policy on mergers and competition was endorsed by many committees on the grounds that foreign firms have become major competitors in many industries. Encouragement of joint R&D ventures was also favored.

Government Procurement

The government has an important role to play in improving the competitiveness of industries in which government or government contractors are major customers, such as aircraft and machine tools. Several panels favored improvement in procurement policies through more coordinated planning and support for modernization among suppliers.

Patent Reform

Several panels favored changes in the patent system to improve the competitiveness of U.S. industry. The recommended changes include a first-to-file system, restoration of time lost due to regulatory requirements, and an effective computer-based patent search and retrieval system. Some panels favored stronger protection against unfair patent use by other countries.

SUMMARY OF SPECIFIC REPORTS

1. Committee for Economic Development—Research and Policy Committee. 1980. Stimulating Technological Progress. New York: Committee for Economic Development.

Background

This was a committee of 28 business and university executives headed by Thomas A. Vanderslice, GTE Corporation. It analyzed how technological progress is affected by certain economic problems, including slow productivity growth, inadequate capital investment, uncertain government regulation, and a complicated patent system.

Major Policy Recommendations

- Tax policy should be changed to increase investment in new plant and equipment, including a more rapid capital recovery allowance and flexible depreciation of fixed R&D assets.

• Regulatory reform should use performance standards, a system of economic incentives and penalties, and appraisal of potential adverse effects on future innovation.

• The patent system could be made more effective through voluntary arbitration, a single court of appeals, and a first-to-file patent system.

• Federal R&D support for basic research in universities should be increased.

2. Committee for Economic Development—Research and Policy Committee. 1983. Productivity Policy: Key to the Nation's Economic Future. New York: Committee for Economic Development.

Background

Under the chairmanship of William F. May, Dean of the New York University Graduate School of Business Administration and Management, 31 executives and consultants undertook a study of the nation's lagging productivity growth in relation to its past record and to the rates of other industrialized economies. The committee emphasized that long-term and, in some cases, fundamental changes are needed in management and labor practices and in public policy areas.

Major Policy Recommendations

• Review the tax code to achieve substantial simplification in the tax structure, reduction in tax preferences, and reduction in marginal tax rates through a broadening of the tax base.

• Adopt a mechanism to adjust the valuation of capital gains for inflation thereby eliminating a major impediment to saving and investment.

• Increase outlays for repairs, modernization, and expansion of the portion of the public infrastructure that contributes to productivity.

• Ensure that investment incentives are neutral among different types of capital assets.

- Increase federal funding of basic research, especially in universities, including payment of the full cost of R&D performed under contract by universities, financing of individual scholars of outstanding ability, and sharing of high cost instruments among universities.
- Introduce a flexible system of depreciation of R&D capital assets by amending laws to permit expensing of R&D structures and equipment.
- Review current antitrust policies and modify any antitrust laws that inhibit productivity growth.
- Regulatory goals should be pursued primarily through market incentives and the use of the bubble and offsets programs; use of regulatory requirements and direct controls should be avoided.
- Every American business should adopt explicit productivity goals and select appropriate techniques to achieve them, including encouraging entrepreneurship within the firm.
- Involve employees and unions in designing and implementing policies to enhance productivity, including compensation systems providing financial incentives for improved productivity.

3. Committee on Technology and International Economic and Trade Issues. 1985. The Competitive Status of the U.S. Civil Aviation Manufacturing Industry. Washington, D.C.: National Academy Press.

Background

The Civil Aviation Industry Panel of the Committee on Technology and International Economic and Trade Issues, headed by Frederick Seitz, included 27 business, labor, and academic experts. They examined key challenges created by a combination of circumstances, including deregulation of airlines, emergence of foreign competitors, internationalization of aircraft manufacture, and growing involvement of foreign governments in the industry.

Major Policy Recommendations

- Monitor trade and encourage compliance with trade agreements.
- Extend measures that would enable aircraft manufacturers to spread the risk in leasing aircraft to domestic and foreign customers.
- Reexamine the lending role of the Export-Import Bank in light of heightened competition.
- Continue the new Export-Import Bank facility to provide medium-term loans for sales of small aircraft.
- Develop mechanisms to ensure an effective industry voice in deliberations on coproduction.
- Reexamine mechanisms for working with civil aircraft manufacturers to ensure that maximum advantage is taken of dual-use capabilities in technology development for design, manufacture, and certification.
- The Department of Defense (DOD) and industry should strengthen the process of coordinated planning for aircraft procurement to reduce, as far as practicable, disruption due to the great cyclicality in production.
- Reexamine the research and technology development activity in support of civil aviation within the National Aeronautics and Space Administration (NASA) in light of new technologies and a changing competitive environment, including expansion of NASA programs on technology validation.

4. Committee on Technology and International Economic and Trade Issues. 1982. The Competitive Status of the U.S. Auto Industry. Washington, D.C.: National Academy Press.

Background

Professor William J. Abernathy of the Harvard University Graduate School of Business Administration headed an 11-person panel of experts from industry, labor, universities, finance, and an official of a major automobile manufacturing company. The panel dealt primarily with trends in cost and technology, recent innovations, and their impact on the industry's relative competitive position. An in-depth analysis of policy options was not carried out. Instead, three main lines of future development were presented, with the broad public policy implications of each line indicated.

Major Policy Recommendations

On the assumption that the current situation reflects a temporary economic misfortune, the panel projects that

- Relaxation of regulation and tax incentives will spur capital investment and allow a viable domestic industry.
- Temporary import quotas will reduce imports, slacken the drive for changes in management, and accelerate construction of Japanese plants in the United States.

On the assumption that the auto industry is maturing and losing competitiveness to low-cost foreign producers, the panel projects that

- Relaxation of regulations and tax incentives will result in capital investment in specialty and high technology models.
- Temporary import quotas will encourage production inefficiencies and high prices, weakening primary demand.

On the assumption that fundamental structural change is taking place with substantial technological change, new products, decline of vertical integration, and advance of specialized producers, the panel projects that

- Tax incentives and deregulation will aid investment in new product development.
- Temporary import quotas will preserve market share in standard models, but reduce the urgency of structural changes.

5. Committee on Technology and International Economic and Trade Issues. 1982. The Competitive Status of the U.S. Electronics Industry. Washington, D.C.: National Academy Press.

Background

The chairman of the 12-person panel of university and industry experts was Professor John G. Linvill of Stanford University. Four sectors of the industry—semiconductors, computers, telecommunications equipment, and consumer electronics—were studied separately, but policy options were outlined for the industry as a whole.

Major Policy Recommendations

Research policy

- Encourage joint research ventures.
- Support basic research.
- Expand incremental R&D funding as well as research grants to universities for projects related to a firm's business.

Capital formation policy

- Create a category in the depreciation system for equipment with high rates of technological obsolescence combined with a reduction in the penalty for taking the investment tax credit over a short time period.
- Increase the first-year depreciation allowance.

Human resource policy

If the industry is provided with appropriate incentives by government, it can take important steps to respond to the shortage of faculty and equipment needed to train additional electrical engineers.

International trade policy

- Clarify and consolidate responsibility for foreign trade policy within one federal agency.
- Review restrictive U.S. laws and regulations and eliminate instances in which they place U.S. firms at a competitive disadvantage.

Electronics industry policy

"The U.S. government, working with industry and universities, must develop a statement of national goals for the electronics industry."

6. Committee on Technology and International Economic and Trade Issues. 1983. The Competitive Status of the U.S. Pharmaceutical Industry. Washington, D.C.: National Academy Press.

Background

A panel of 12 industry, financial, and university experts, headed by Charles C. Edwards, President, Scripps Clinic and Research Foundation, examined the relative decline of the U.S. pharmaceutical industry despite its continued expansion of output.

Major Policy Recommendations

- The Food and Drug Administration (FDA) prohibition of exports of unapproved new drugs should be revised to permit domestic production for shipment abroad.
- Restore the patent time lost as a result of FDA regulatory requirements.
- Examine antitrust policy to determine whether it discourages mergers that could make U.S. companies more competitive in world markets.
- Expand research tax credits to include research-related expenditures not now eligible for the investment tax credit.
- Allocate R&D expenditures incurred in the United States solely to the U.S. income of the taxpayer.
- Study the impact of product liability in the pharmaceutical industry in an attempt to reduce this disincentive to research.
- Implement the recommendations of the Commission on the Federal Drug Approval Process to expedite the review process without reducing public health protection.

7. Committee on Technology and International Economic and Trade Issues. 1983. The Competitive Status of the U.S. Fibers, Textiles, and Apparel Complex. Washington, D.C.: National Academy Press.

Background

An 11-person panel of experts from industry, labor, universities, and the trade press, headed by Dean W. Denney Freeston, Jr., of the Georgia Institute of Technology College of Engineering, assessed the future of international competitiveness of the U.S. textile complex, including fibers, fabrics, and end users—apparel, home furnishings, and industrial. One of every eight factory workers is employed in this textile sector. This panel concentrated on incremental changes in existing policies rather than on sweeping changes.

Major Policy Recommendations

Trade

- Enforce existing trade mechanisms more rigorously.
- Tighten controls and speed response to changes in market conditions and import surges.
- Seek reductions in tariff and nontariff barriers in other countries.
- Change to a system of granting licenses to U.S. importers instead of foreign exporters.

Technology

- Examine government-sponsored collaborative R&D projects in apparel manufacturing, involving fiber, fabric, and equipment industries.
- Place greater emphasis on policies affecting equipment utilization, such as more favorable tax incentives for the use of experimental equipment.

Education and training

- Increase financial support for expansion of technical and supervisory training programs and applied engineering programs.
- Increase emphasis on and programs in business schools on developing needed skills for middle management and small businesses.

8. Committee on Technology and International Economic and Trade Issues. 1983. The Competitive Status of the U.S. Machine Tool Industry. Washington, D.C.: National Academy Press.

Background

A panel of 11 industry, labor, financial, and university experts, headed by E. Raymond McClure of Lawrence Livermore National Laboratory, examined the international competitive position of the U.S. machine tool industry, emphasizing the areas of trade and technology. The panel considered a wide set of policy options, specifically rejecting any large-scale government subsidies, protectionism, and major countercyclical purchases of machine tools by DOD. It concentrated on the removal of obstacles posed by existing policies rather than on crafting dramatically new ones.

Major Policy Recommendations

Trade policy

Policies are needed to maintain at least a 15 percent surplus of machine tool exports over imports, with exports, chiefly high technology, restored to levels previously attained. Such policies include

- The development of a national export policy.
- An increase in the sensitivity of the Department of Commerce to the export needs of small business.

- The alteration of tax policies that now discourage foreign trade and encourage foreign investment.
- The review of the loan policies of the Export-Import Bank as they affect small business.
- The review of restrictions applied to exports to the Soviet Union and the Eastern bloc that machine tool builders deem too extensive.

Manpower policy

- Recruit 300 new Ph.D.s per year in manufacturing engineering to alleviate personnel shortages.
- Develop training programs with industry and government backing to train machinists and toolmakers.

Capital formation

- Fashion monetary policy to stabilize interest rates to encourage long-term capital investment.
- Clarify guidelines to industry for antitrust enforcement and launch an analysis of the effects of antitrust prohibitions against domestic mergers and market concentrations on international competitiveness.

Machine tool technology

- Emphasize manufacturing technology in federal R&D funding to support generic and high-risk research.
- Allow R&D tax write-offs to be spread over several years.

Regulatory environment

- Review many specific Occupational Safety and Health Administration (OSHA) regulations for their impact on competitiveness.
- Adopt on a state-by-state basis the Department of Commerce's Model Product Liability Law.

9. *Committee on Technology and International Economic and Trade Issues. 1985. The Competitive Status of the U.S. Steel Industry. Washington, D.C.: National Academy Press.*

Background

Under the chairmanship of Bruce S. Old of Bruce S. Old Associates, Inc., the panel of 16 experts from producers and customers, academia, consultants, and labor considered the problems of the mature and declining U.S. steel industry in the context of the depressed world steel market. The areas of major policy alternatives (not recommendations) available to government, management, and labor decision makers cover trade, environment, antitrust, and technology.

Major Policy Options

Trade

The panel considered the benefits and disadvantages of antidumping suits, countervailing duties, tariffs, trigger price mechanisms, and quantitative limitations, including the current system of negotiated voluntary quotas.

Environmental policy

The panel found that controls will not pose a major problem if only current environmental requirements are enforced.

Antitrust

A consistent, flexible, and less restrictive policy toward mergers and acquisitions, joint ventures, and jointly sponsored R&D might not adversely affect steel consumers, if imports and the large number of domestic producers continue.

Technology

Government should continue funding basic research on materials; cooperative research by steel companies on large costly projects was also endorsed.

10. Committee on the Machine Tool Industry. 1983. The U.S. Machine Tool Industry and the Defense Industrial Base. Washington, D.C.: National Academy Press.

Background

The study was made in 1983 by the Committee on the Machine Tool Industry, an 18-member group of academics, bankers, and experts from industries that produce and use machine tools. The committee was headed by James E. Ashton, Vice President, Rockwell International. The study was undertaken for DOD to assess the international competitiveness of the domestic machine tool industry, study its current and expected responsiveness to defense needs, and recommend policies for DOD and others to ensure access to sufficient domestic machine tool capacity and capability.

Major Policy Recommendations

DOD actions

- Modernize the defense industrial base through increased and stable funding of the Industrial Modernization Incentives Program and the Manufacturing Technology programs.
- Stress productivity improvement incentives within the machine tool industry.
- Simplify contracting procedures to encourage individual machine tool firms to bid directly for government contracts.
- Improve information flows between DOD research programs and equipment suppliers.
- Require long-term guarantees for production equipment maintenance in defense contracts.

- Study the effects of consolidation, acquisitions, and joint ventures on the strength of the U.S. industry in competition with foreign producers.

Other government agency actions

- Promote a machine tool export program through the Department of Commerce.
- Reduce barriers to the export of machine tools to Eastern bloc markets.
- Make an inventory of federal programs that are aimed at the problems of manufacturing productivity to gain better coordination and simplification.

Machine tool industry actions

- Aggressively apply advanced equipment and processes in machine tool production.
- Actively search for new technology.
- Increase investments in long-term competitive strategies rather than responding only to short-term economic considerations.
- Participate in joint R&D efforts.
- Expand information programs to inform machine tool builders of DOD programs.

11. Committee on the CAD/CAM Interface. 1984. Computer Integration of Engineering Design and Production: A National Opportunity. Washington, D.C.: National Academy Press.

Background

In response to a request from NASA, the Committee on the CAD/CAM Interface, composed of 12 experts from industry and academia, was organized by the Manufacturing Studies Board to recommend ways to improve the interaction between the engineering design of a product and its production. Professor Arthur R. Thomson chaired the committee. The committee visited five companies with experience in integration and reviewed three major federal programs.

Major Policy Recommendations

- A strategy of computer-integrated manufacturing (CIM) should be adopted by NASA for its manned space station program.
- Consortia should be formed by groups of companies to pursue research and other projects in CIM not readily undertaken by individual companies.
- Existing knowledge of CIM technology should be compiled by the Computer and Automated Systems Association and made available to industry, universities, and government agencies.
- Research should continue to be undertaken by the federal government to resolve fundamental technical issues related to CIM.
- Federal agencies that purchase manufactured goods should accept digital data sets compatible with the Initial Graphic Exchange Standard rather than requiring conventional drawings as a deliverable item under contracts.
- Manufacturing companies considering investment in product design or manufacturing process technology should consider CIM.

12. Computer Conferences on Productivity. 1983. A Final Report for the White House Conference on Productivity. Houston: American Productivity Center.

Background

The American Productivity Center brought together 175 senior-level leaders from business, labor, academia, and government through a computer conferencing system over a four-month period in 1983. These leaders exchanged information and debated recommendations in areas involving productivity and work quality for submission to the White House Conference on Productivity.

The consensus reached in several areas pertains to the manufacturing sector.

Major Policy Recommendations

- A less authoritarian and more interactive style of management should be followed at all levels.
- Organized labor should accept greater responsibility for the competitiveness of its employing firms.
- Government is responsible for moderating the human impact of the competitive process.
- Quality awareness should be raised through campaigns that would include private cooperation with schools to spread the concept of quality and national awards for contributions to improved quality.
- Reward systems for productivity improvement should be initiated, including sharing business information with employees, participative work practices, pay for performance, and better measurement of productivity.
- Strong programs of education and training are needed to enhance management skills in the development and use of new technology.
- Government procurement practices should be changed to include provisions for sharing increases in risk and reductions in cost due to implementation of new technology.

13. White House Conference on Productivity. 1984. Productivity Growth: A Better Life for America. Washington, D.C.: U.S. Government Printing Office.

Background

On October 25, 1982, the President signed legislation calling for a White House Conference on Productivity to develop recommendations for stimulating productivity growth in the United States. During the following 11 months, preparatory conferences were held at four universities across the country, each dealing with a single topic. The White House Conference was held in Washington, D.C., on September 21-23, 1983, and was chaired by L. William Seidman and William Simon. About 1,000 persons attended to discuss the findings and recommendations of the preparatory conferences and new suggestions for action by the public and private sectors.

Major Policy Recommendations

Government actions

• Increase public recognition and acceptance of improving productivity growth as a national goal and as the means of raising our standard of living, including

— consistently evaluating regulations and laws in terms of effects on productivity;
— creating a National Medal for Productivity and Quality Achievement; and
— emphasizing better techniques of measuring productivity.

• Maintain a stable noninflationary economic environment and reduce the government's consumption of national resources.
• Develop a specific plan for fundamental tax reform, with improving productivity as a standard for evaluating tax reform proposals.
• Change or repeal laws that impose impediments to productivity growth, including

— performance standards in environmental regulation;
— cost-benefit analysis and market-based incentives in health and safety regulation;
— protection of patents against unfair use by other countries, especially in areas of computer software and chip technology;
— making joint ventures, including joint R&D ventures, a more effective means of meeting world competition; and

— removal of restraints on competition in the energy, communications, transportation, and financial service industries.

Private sector actions

- Focus more attention on improving technology, quality, and information resources.
- Employ creative, innovative work practices to use more fully the knowledge and talent of employees.
- Establish productivity measures and improvement goals, especially for information and service workers.
- Promote labor-management cooperation to consider workplace problems such as plant closings, training, restrictive practices, and employment security.

14. U.S. Congress, Office of Technology Assessment. 1984. Computerized Manufacturing Automation: Employment, Education, and the Workplace. Washington, D.C.: U.S. Government Printing Office.

Background

The Office of Technology Assessment (OTA), with a 24-person advisory panel and three workshops, conducted an in-depth study of programmable automation, covering computer-aided design, computer-aided manufacturing, and computer-aided techniques for management. The findings covered the state of the art, the impact on employment and occupations, health and safety effects, education, training and retraining issues, the automation industries, R&D, and programs abroad. Instead of specific recommendations, the OTA proposed a series of policy options for congressional consideration under four broad headings.

Major Policy Options

Technology development and diffusion

- Increase funds for R&D on automation, especially for longer-term generic research in nonmilitary applications.
- Facilitate standard-setting as a means of increasing the ease of use and encouraging the application of automation technologies.
- Redress the historical U.S. inattention to manufacturing processes, organization, and management through support of engineering education and some form of manufacturing institute and clearinghouse.

Employment programs

- Establish job creation programs for production of public goods and services, from highway building to child care.
- Expand programs for disseminating labor market information.
- Expand programs for assisting displaced workers, including advance notice and financial incentives to relocate personnel, either in or outside the firm.

Work environment

Increase oversight of and research on the workplace effects of automation through OSHA and the National Institute for Occupational Safety and Health.

Education, training, and retraining

- Increase support for facilities, equipment, and qualified instructors at colleges, universities, and vocational schools.
- Encourage curriculum development geared to the development of automation-related skills.
- Encourage renewed emphasis on basic skills in reading, math, and science as well as problem-solving skills.

• Encourage individual participation in instruction related to automation and development of industry in-house programs.

15. Advisory Committee on Industrial Innovation. 1979. Final Report. Washington, D.C.: Department of Commerce.

Background

In 1978, President Carter called for an extensive review of government policies on industrial innovation. Under the guidance of the Department of Commerce, more than 150 senior representatives from industry, labor, academia, science, and public interest groups participated in meetings held by the Advisory Committee on Industrial Innovation. These deliberations covered the broad issues of economic and trade policy, environmental, health, and safety regulations, regulation of industry structure and competition, federal patent and information policy, federal procurement policy, and direct federal support of R&D. Industrial members of the advisory committee issued 10 separate reports covering the effects of federal policies in these areas on industrial innovation and specific recommendations for change. Separate reports covering all issues were produced from the perspectives of small business, labor, and public interest groups.

Major Policy Recommendations

Since the specific recommendations are too numerous to list, only the major general ones are highlighted, under five broad headings.

Economic and trade policy

• Revise tax laws to eliminate disincentives to overall investment and R&D.

• Reduce disincentives to savings to help alleviate shortages of venture capital.

• Maintain strong foreign competition as a spur to innovation.

Environmental, health, and safety regulation

• Improve the regulatory process through analysis of risks, costs, and benefits, reducing uncertainty of content and timing, etc.

• Maintain a high level of competence in the regulatory agencies.

• Avoid use of mandatory controls, including methods specification.

Regulation of industry structure and competition

• Encourage joint or cooperative research, even among large competitors in some cases.

• Consider general economic health in enforcement of antitrust policy.

Patent and information policy

• Complete development of an effective computer-based patent search and retrieval system.

• Establish a policy of convenient access to all information created and collected by the government, except for confidential and classified materials.

Direct federal support of research and development

• Support a substantial increase in the coupling of university research and industrial needs.

• Support R&D and the dissemination of new technology generic to process or product innovation in a wide array of industries.

16. The President's Commission on Industrial Competitiveness. 1985. Global Competition: The New Reality. 2 vols. Washington, D.C.: U.S. Government Printing Office.

Background

The President formed the 30-person Commission on Industrial Competitiveness in 1983 to study ways of improving the competitiveness of American industry. The commission, under the chairmanship of John A. Young, President, Hewlett-Packard Corporation, included high officials of private industry, unions, universities, and banks. The commission's recommendations are reported under four headings.

Major Policy Recommendations

Research and development and manufacturing

- Create a federal Department of Science and Technology.
- Increase tax incentives for R&D.
- Remove antitrust barriers to joint R&D.
- Commercialize new technologies through improved manufacturing processes.
- Strengthen protection of intellectual property rights.

Capital resources

- Reduce the federal deficit.
- Restructure the tax system.
- Pursue a stable monetary policy.
- Remove barriers to the efficient flow of capital.

Human resources

- Increase effective dialogue among government, industry, and labor through existing advisory committees and in other ways.
- Encourage labor-management cooperation through presidential recognition of good cases.
- Strengthen employee incentives through use of plans linking pay and performance.
- Encourage programs to assist in the reemployment of displaced workers.
- Improve engineering education and business schools.

- Establish partnerships in education to counter high dropout rates and continue support of research programs in educational software for computer technology.

International trade

- Improve trade and investment policy through the establishment of a Department of Trade.
- Review U.S. trade law to facilitate industrial adjustment to increased global competition.
- Reform U.S. antitrust policy to recognize the potential efficiency gains from business combinations and the reality of global competition.
- Review export policies, including renewal of the Export Administration Act, negotiate agreements to minimize the impact of national security export controls on competitiveness, and improve the functioning of export expansion, trade information, and export financing programs.

17. Work in America Institute. 1984. Employee Security in a Free Economy. New York: Pergamon Press.

Background

The Work in America Institute, a nonprofit, nonpartisan organization, conducted an investigation of workable alternatives for achieving greater employment stability as a means of spurring company prosperity. A national advisory committee of 38 experts from business, labor, academia, and consultancy provided guidance to the study. The study was directed by Jerome M. Rosow, President of the Work in America Institute, and Robert Zager, Vice President for policy and technical studies. The specific recommendations fall under seven major headings.

Major Policy Recommendations

- Employers should consider a commitment to employment security to the full extent in circumstances that are under their control and to the extent that they can in circumstances beyond their control.

- Employers should institute a system of advance planning of human resource needs and shifts, in consultation with union officials.
- Employers should adhere to lean staffing standards to avoid layoffs due to imbalances and should adopt production, marketing, and financial policies that minimize sudden changes in the size of the protected work force.
- In response to economic declines, employers should defer layoffs as long as possible while taking advantage of attrition, release temporary employees, increase training, and, as a final resort, use work sharing.
- Employers, in the event of dismissal of permanent employees, should actively help them find suitable work elsewhere, including outplacement services, retraining, and pension portability.
- Regional groups of employers and unions should form alliances and organize computer-based job clearinghouses and retraining, education, and job creation programs.
- Government should play a supportive role, providing incentives to employers who are committed to employment security, requiring protection to provide an approximate measure of employment security, and enacting short-time compensation regulation.

Bibliography

Abegglen, James C., and George Stalk, Jr. 1985. Kaisha: The Japanese Corporation. New York: Basic Books, Inc.

Abernathy, William J., Kim B. Clark, and A. M. Kantrow. 1983. Industrial Renaissance: Producing a Competitive Future for America. New York: Basic Books.

Advisory Committee on Industrial Innovation. 1979. Final Report. Washington, D.C.: U.S. Department of Commerce.

Amrine, Harold T., John A. Ritchey, and Oliver S. Hulley. 1982. Manufacturing Organization and Management. Englewood Cliffs, N.J.: Prentice-Hall.

Baumol, William J., and Kenneth McLennan, eds. 1985. Productivity Growth and U.S. Competitiveness. New York: Oxford University Press.

Bell, Daniel. 1976. The Coming of Post-Industrial Society. New York: Basic Books.

Bluestone, Barry, and Bennett Harrison. 1982. The Deindustrialization of America. New York: Basic Books.

Blumenthal, Marjory, and Jim Dray. 1985. The Automated Factory: Vision and Reality. Technology Review 88(1):28-37.

Booz-Allen and Hamilton, Inc. 1985. Manufacturing Issues 1985. New York: Booz-Allen and Hamilton, Inc.

Boskin, Michael J. 1985. The Impact of the 1981-1982 Investment

Incentives on Business Fixed Investment. Washington, D.C.: National Chamber Foundation.

Bureau of Labor Statistics. 1985. Trends in Manufacturing: A Chartbook. Washington, D.C.: U.S. Government Printing Office.

Childs, James J. 1982. Principles of Numerical Control. New York: Industrial Press, Inc.

Clark, Kim B., Robert H. Hayes, and Christopher Lorenz, eds. 1985. The Uneasy Alliance: Managing the Productivity Dilemma. Boston, Mass.: Harvard Business School Press.

Committee for Economic Development—Research and Policy Committee. 1980. Stimulating Technological Progress. New York: Committee for Economic Development.

Committee for Economic Development—Research and Policy Committee. 1983. Productivity Policy: Key to the Nation's Economic Future. New York: Committee for Economic Development.

Committee on Army Robotics and Artificial Intelligence. 1983. Applications of Robotics and Artificial Intelligence to Reduce Risk and Improve Effectiveness. Washington, D.C.: National Academy Press.

Committee on the CAD/CAM Interface. 1984. Computer Integration of Engineering Design and Production: A National Opportunity. Washington, D.C.: National Academy Press.

Committee on the Effective Implementation of Advanced Manufacturing Technology. 1986. Human Resource Practices for Implementing Advanced Manufacturing Technology. Washington, D.C.: National Academy Press.

Committee on the Evolution of Work. 1985. The Changing Situation of Workers and Their Unions. Washington, D.C.: AFL-CIO.

Committee on the Machine Tool Industry. 1983. The U.S. Machine Tool Industry and the Defense Industrial Base. Washington, D.C.: National Academy Press.

Committee on Materials Information Used in Computerized Design and Manufacturing Processes. 1983. Materials Properties Data Management—Approaches to a Critical National Need. Washington, D.C.: National Academy Press.

Committee on Science, Engineering, and Public Policy. 1985. New Pathways in Science and Technology: Collected Research Briefings 1982-1984. New York: Vintage Books.

Committee on the Status of High-Technology Ceramics in Japan. 1984. High-Technology Ceramics in Japan. Washington, D.C.: National Academy Press.

Committee on Technology and International Economic and Trade Issues. 1982. The Competitive Status of the U.S. Auto Industry. Washington, D.C.: National Academy Press.

Committee on Technology and International Economic and Trade Issues. 1982. The Competitive Status of the U.S. Electronics Industry. Washington, D.C.: National Academy Press.

Committee on Technology and International Economic and Trade Issues. 1983. The Competitive Status of the U.S. Fibers, Textiles, and Apparel Complex. Washington, D.C.: National Academy Press.

Committee on Technology and International Economic and Trade Issues. 1983. The Competitive Status of the U.S. Machine Tool Industry. Washington, D.C.: National Academy Press.

Committee on Technology and International Economic and Trade Issues. 1983. The Competitive Status of the U.S. Pharmaceutical Industry. Washington, D.C.: National Academy Press.

Committee on Technology and International Economic and Trade Issues. 1985. The Competitive Status of the U.S. Civil Aviation Manufacturing Industry. Washington, D.C.: National Academy Press.

Committee on Technology and International Economic and Trade Issues. 1985. The Competitive Status of the U.S. Steel Industry. Washington, D.C.: National Academy Press.

Committee on U.S. Shipbuilding Technology. 1984. Toward More Productive Naval Shipbuilding. Washington, D.C.: National Academy Press.

Computer Conferences on Productivity. 1983. A Final Report for the White House Conference on Productivity. Houston, Tex.: American Productivity Center.

Cyert, Richard, M. 1985. The Plight of Manufacturing: What Can Be Done? Issues in Science and Technology (Summer):87-100.

Dewar, Margaret E., ed. 1982. Industry Vitalization: Toward a National Industrial Policy. New York: Pergamon Press.

Eckstein, Otto, Christopher Caton, Roger Brinner, and Peter Duprey. 1984. The DRI Report on U.S. Manufacturing Industries. New York: McGraw-Hill Book Company.

Ehner, William J., and F.R. Bax. 1982. Evaluating the Factory of the Future. Manufacturing Productivity Frontiers 6(4):15-22.

Ferdows, Kasra, Jeffrey G. Miller, Jinchiro Nakane, and Thomas E. Vollmann. 1985. Evolving Manufacturing Strategies in Europe, Japan, and North America. Boston, Mass.: Boston University.

Foster, Richard N. 1986. Innovation: The Attacker's Advantage. New York: Summit Books.

Gold, Bela. 1979. Productivity, Technology, and Capital: Economic Analysis, Managerial Strategies, and Governmental Policies. Lexington, Mass.: D.C. Heath-Lexington Books.

Gold, Bela, William S. Pierce, Gerhard Rosegger, and Mark Perlman. 1984. Technological Progress and Industrial Leadership: The Growth of the U.S. Steel Industry, 1900-1970. Lexington, Mass.: D.C. Heath-Lexington Books.

Goldrath, E.M., and J. Cox. 1984. The Goal—Excellence in Manufacturing. Croton-on-Hudson, N.Y.: North River Press.

Groover, Mikell P. 1980. Automation, Production Systems, and Computer-Aided Manufacturing. New York: Prentice-Hall.

Grunwald, Joseph, and Kenneth Flamm. 1985. The Global Factory: Foreign Assembly in International Trade. Washington, D.C.: The Brookings Institution.

Harrington, Joseph, Jr. 1984. Understanding the Manufacturing Process: Key to Successful CAD/CAM Implementation. New York: Marcel Dekker, Inc.

Hatsopoulos, George N. 1983. High Cost of American Industry. American Business Conference, Inc. Thermo Electron Corporation. April 26, 1983, p. 118.

Hayes, Robert H., and Steven C. Wheelwright. 1984. Restoring Our Competitive Edge: Competing Through Manufacturing. New York: John Wiley and Sons.

Hounshell, David A. 1984. From the American System to Mass Production, 1800-1932. Baltimore, Md.: Johns Hopkins University Press.

Illinois Institute of Technology Research Institute. 1984. Business Assessment of the Manufacturing Automation Industry (Machine Tools, CAD/CAM, Robotics, FMS). New York: Industrial Research Institute, Inc.

Japan Productivity Center. 1984. Strategies for Productivity: International Perspectives. New York: Unipub.

Johnson, Chalmers, ed. 1984. The Industrial Policy Debate. San Francisco, Calif.: Institute for Contemporary Studies Press.

Jurgen, Ronald K., ed. 1983. Data-Driven Automation. IEEE Spectrum (May):34-96.

Kaplan, Robert S. 1984. Yesterday's Accounting Undermines Production. Harvard Business Review 62(4):95-101.

Kendrick, John W., ed. 1984. International Comparisons of Productivity and Causes of the Slowdown. Cambridge, Mass.: Ballinger Publishing Company.

Kozikowski, Walter F. 1985. U.S. Manufacturers' Five-Year Industrial Automation Plans for Automation Machinery and Plant Communication Systems. Washington, D.C.: National Electrical Manufacturers Association.

Lawrence, Robert Z. 1985. Can America Compete? Washington, D.C.: The Brookings Institution.

Leontief, Wassily, and Faye Duchin. 1984. The Impacts of Automation on Employment, 1963-2000. New York: New York University Institute for Economic Analysis.

Lodge, George, and Bruce Scott, eds. 1985. U.S. Competitiveness in the World Economy. Cambridge, Mass.: Harvard Business School Press.

Lund, Robert T., and John A. Hansen. 1986. Keeping America at Work: Strategies for Employing the New Technologies. New York: John Wiley and Sons.

Magaziner, Ira C., and Robert B. Reich. 1983. Minding America's Business. New York: Vintage Books.

McClure, E. Raymond. 1985. Ultraprecision Machining and the Niche of Accuracy. CIM (September/October):16-20.

McKinsey and Company. 1983. Computer Integrated Manufacturing. Cleveland: McKinsey and Company.

Merris, Dora. 1985. Post-Automation Society. The Journal of the Institute for Socioeconomic Studies 10(2):53-66.

Millar, Gordon H. 1985. Computer Integrated Manufacturing and Its Impact on Engineering and the Profession. Warrendale, Pa.: Society of Automotive Engineers, Inc.

Miller, Jeffrey G., and Thomas E. Vollmann. The Hidden Factory. Harvard Business Review 63(5):142-150.

Moxley, David. 1985. The U.S. Trade Deficit: Impact and Implications. New York: Touche Ross and Company.

National Academy of Engineering. 1983. U.S. Leadership in Manufacturing. Washington, D.C.: National Academy Press.

National Academy of Engineering. 1984. The Long-Term Impact of Technology on Employment and Unemployment. Washington, D.C.: National Academy Press.

National Academy of Engineering. 1985. Education for the Manufacturing World of the Future. Washington, D.C.: National Academy Press.

Nelson, Richard R., and Richard N. Langlois. 1985. Industrial Innovation Policy: Lessons from American History. Science 219:814-818.

New York Stock Exchange, Office of Economic Research. 1984. U.S. International Competitiveness: Perception and Reality. New York: New York Stock Exchange.

Noble, David F. 1984. Forces of Production: A Social History of Industrial Automation. New York: Knopf.

Ohmae, Kenichi. 1982. The Mind of the Strategist: The Art of Japanese Business. New York: McGraw-Hill.

Ohmae, Kenichi. 1985. Triad Power: The Coming Shape of Global Competition. New York: The Free Press.

Ohri, Rajni Bonnie. 1985. U.S. Manufacturing Is Alive and Well. Manufacturing Productivity Frontiers (October):1-7.

Pollard, Sidney. 1982. The Wasting of the British Economy: British Economic Policy 1945 to the Present. New York: St. Martin's Press.

The President's Commission on Industrial Competitiveness. 1985. Global Competition: The New Reality. 2 vols. Washington, D.C.: U.S. Government Printing Office.

Ranky, Paul G. 1983. The Design and Operation of FMS: Manufacturing Systems. Kempston, UK: IFS Publications, Ltd.

Reich, Robert. 1983. The Next American Frontier. New York: Times Books.

Research Briefing Panel on Ceramics and Ceramic Composites. 1985. Research Briefings 1985. Washington, D.C.: National Academy Press.

Schonberger, Richard J. 1982. Japanese Manufacturing Techniques: Nine Hidden Lessons in Simplicity. New York: The Free Press.

Shaiken, Harley. 1985. The Automated Factory: The View from the Shop Floor. Technology Review 88(1):16-27.

Shaiken, Harley. 1985. Work Transformed: Automation and Labor in the Computer Age. New York: Holt, Rinehart, and Winston.

Shunk, Dan. 1983. Integrated Cellular Manufacturing. Paper presented at the WESTEC conference, Los Angeles, Calif., March 1983.

Skinner, Wickham. 1985. Manufacturing—The Formidable Competitive Weapon. New York: John Wiley and Sons.

Slade, Bernard N., and Raj Mohindra. 1985. Winning the Productivity Race. Lexington, Mass.: D.C. Heath-Lexington Books.

Smith, Donald N., and Peter Heyler, Jr. 1985. U.S. Industrial Robot Forecast and Trends: A Second Edition Delphi Study. Dearborn, Mich.: Society of Manufacturing Engineers.

Solberg, James J., David C. Anderson, Moshe M. Barash, and Richard P. Paul. 1985. Factories of the Future: Defining the Target. West Lafayette, Ind.: Purdue University Computer Integrated Design, Manufacturing, and Automation Center.

Taniguchi, Norio. 1983. Current Status in, and Future Trend of, Ultraprecision Machining and Ultrafine Materials Processing. Annals of the International Society of Production Research (CIRP) 32:573-582.

Thompson, Brian. 1985. Fixturing: The Next Frontier in the Evolution of Flexible Manufacturing Cells. CIM (March/April): 10-13.

Thurow, Lester C. 1985. The Zero-Sum Solution: Building a World Class American Economy. New York: Simon and Schuster.

U.S. Congress, Office of Technology Assessment. 1980. Technology and Steel Industry Competitiveness. Washington, D.C.: U.S. Government Printing Office.

U.S. Congress, Office of Technology Assessment. 1980. Technology and Structural Unemployment: Reemploying Displaced Adults. Washington, D.C.: U.S. Government Printing Office.

U.S. Congress, Office of Technology Assessment. 1981. U.S. Industrial Competitiveness: A Comparison of Steel, Electronics, and Automobiles. Washington, D.C.: U.S. Government Printing Office.

U.S. Congress, Office of Technology Assessment. 1982. Exploratory Workshop on the Social Impacts of Robotics: Summary and Issues. February 1982. Washington, D.C.: U.S. Government Printing Office.

U.S. Congress, Office of Technology Assessment. 1983. International Competitiveness in Electronics. Washington, D.C.: U.S. Government Printing Office.

U.S. Congress, Office of Technology Assessment. 1984. Computerized Manufacturing Automation: Employment, Education, and the Workplace. Washington, D.C.: U.S. Government Printing Office.

U.S. Congress, Office of Technology Assessment. 1985. Information Technology R&D: Critical Trends and Issues. Washington, D.C.: U.S. Government Printing Office.

U.S. General Accounting Office. 1983. Federal Efforts Regarding Automated Manufacturing Need Stronger Leadership. Gaithersburg, Md.: General Accounting Office.

U.S. International Trade Administration. 1983. A Historical Comparison of the Cost of Financial Capital in France, the Federal Republic of Germany, Japan, and the United States. Washington, D.C.: U.S. Department of Commerce.

The Wall Street Journal. September 16, 1985. A Special Report: Technology in the Workplace, pp. C1-C100.

White House Conference on Productivity. 1984. Productivity Growth: A Better Life for America. Washington, D.C.: U.S. Government Printing Office.

Work in America Institute. 1984. Employee Security in a Free Economy. New York: Pergamon Press.

Zucker, Seymour, Claudia H. Deutsch, John Hoerr, Norman Jones, John S. Pearson, and James C. Cooper. 1982. Re-industrialization of America. New York: McGraw-Hill.

Zysman, John, and Laura Tyson, eds. 1983. American Industry in International Competition: Government Policy and Corporate Strategies. Ithaca, N.Y.: Cornell University Press.

Index

D

E

K

L

M

Q

R

S

T

U

V

W

X